IoT Technologies for Renewable Energy Generation

This book provides an understanding of the significant applications of IoT technologies in renewable energy. It explores their integration with various renewable energy sources, including solar, wind, bioenergy, piezoelectric materials, and solutions addressing water shortage challenges, highlighting advancements and practical implementations in each area. A comparative analysis of recent techniques is presented to evaluate the efficiency of IoT-based renewable energy systems. This book also focuses on the economic feasibility of IoT-enabled renewable energy solutions.

Features:

- Addresses multiple renewable energy technologies with IoT-based monitoring and control.
- Explores solar, wind, and hybrid energy systems, including energy storage and microgrid integration.
- Investigates renewable energy microgrid systems.
- Focuses on climate goals and the Sustainable Development Goals.
- Includes discussions of marine renewable energies and offshore wind turbine generation.

This book is aimed at graduate students and researchers in renewable energy engineering, IoT and embedded systems, and electrical engineering.

IoT Technologies for Renewable Energy Generation

Assessment, Monitoring, and Control of Resources

Edited by
Tien Anh Tran, Victor Bolbot, and Ivan CK Tam

CRC Press
Taylor & Francis Group
Boca Raton London New York

CRC Press is an imprint of the
Taylor & Francis Group, an **informa** business

Designed cover image: Getty Images

First edition published 2027
by CRC Press
2385 NW Executive Center Drive, Suite 320, Boca Raton FL 33431

and by CRC Press
4 Park Square, Milton Park, Abingdon, Oxon, OX14 4RN

CRC Press is an imprint of Taylor & Francis Group, LLC

ISBN: 9781041136668 (hbk)
ISBN: 9781041136880 (pbk)
ISBN: 9781003671039 (ebk)

DOI: 10.1201/9781003671039

Typeset in Times
by codeMantra

Contents

Editors' Biography

Tien Anh Tran (IEEE Member, IMarEST Member) is a Research Fellow at the Department of Naval Architecture and Ocean Engineering, Seoul National University, Seoul city, South Korea & Researcher at the Department of Electrical Engineering, University of Malta, Malta.

He is an Assistant Professor at the Department of Marine Engineering, Vietnam Maritime University, Haiphong City, Vietnam. Additionally, he is also an Honorary Professor at School of Computing Science and Engineering, Galgotias University, India. He received B.Eng and M.Sc degrees from Vietnam Maritime University, Vietnam, in 2011 and 2014, respectively. He obtained a PhD degree from Wuhan University of Technology, Wuhan, China, in 2018. He is an Editor/Guest Editor for the reputable journals indexed in SCI/SCIE. In 2015, he was awarded the Chinese Government Scholarship (CSC) for the full funding of Doctor of Philosophy (PhD) program in China. In 2019, He was awarded the NEPTUNE prize for outstanding researchers by Vietnam Maritime University. In 2022, he was one of five outstanding scientists in Vietnam to be nominated for Ta Quang Buu prize by the National Foundation for Science & Technology Development (NAFOSTED). Additionally, he had been selected and awarded the full scholarship for the Postdoctoral Fellowship Program of National Research Foundation (NRF) for Foreign Researchers by the Government of South Korea.

Victor Bolbot is a Naval Architect and Marine Engineer specializing in maritime safety, autonomous ships, and sustainable marine systems. He is a Postdoctoral Researcher at Aalto University, Finland, where he contributes to major international projects on autonomous ship design, cyber-risk assessment, accident analysis, and the sustainability of Arctic operations. He has participated in over 25 research projects, including the EU Horizon AUTOSHIP initiative and collaborations with ABB, VTT, DNV, and Wärtsilä. He received an M.Eng. degree from the National Technical University of Athens and a PhD from the University of Strathclyde, where he studied safety in marine cyber-physical systems. His earlier industry experience includes serving as Performance Engineer and TMSA coordinator at Pantheon Tankers Management Ltd. He has led/coauthored 35 journal papers, 33 conference papers, and 12 technical reports, accumulating more than

1,600 citations (h-index 21). His work spans ship systems safety, maritime cybersecurity, autonomous navigation, AIS data analytics, and energy systems. He has received several awards, including distinctions from DNV, RINA, IMarEST, and TRA VISIONS 2020. He has contributed to teaching on five university courses and supervised 20 students across Bachelor, Master, PhD, and D.Sc. levels. He maintains active collaborations in over 12 countries and has lived and worked in Greece, the UK, Finland, and Malaysia.

Ivan CK Tam is an Associate Professor at Newcastle University in the UK with a wealth of experience in multidisciplinary research and a strong track record of leading innovative projects. His research interests are in the clean fuel combustion process, exhaust emission control, energy management, and renewable energy technology. He has successfully applied cryogenic technology and organic Rankine cycles to improve performance in marine engines. He is currently developing a novel wet scrubber technology in the mitigation of gas emissions, including carbon dioxide with his research team. He has a good connection with engineering companies through the network of learned societies and consultancy work with local industries in Singapore. He is a Fellow member, Assessor of Registration, the Honorary Advisor, and Immediate Past Chairman of the Joint Branch (IMarEST & RINA). He received the President's Commendation Award from IMarEST in 2016. He is the Director of Degree Programmes in Marine Technology (Marine Engineering, Naval Architecture & Offshore Engineering) at the university. He is currently holding an appointment as the Director of Innovation & Engagement. A key aim of this new office is to increase engagement with alumni and partnership with industry. He is the Guest Editor for the "Energies" journal special issues "Thermal Systems" and "Advanced Thermal Systems" (IF = 3.252, Citescore = 5.0).

Contributors

Fredah Batale Abifarin, Department of Metallurgical and Materials Engineering, Ahmadu Bello University, Zaria, Nigeria.

Johnson Kehinde Abifarin, School of Engineering, College of Systems and Society, Australian National University, Canberra, Australia.

David Oyeyemi Aboyeji, Department of Microbiology, Federal University of Technology, Minna, Nigeria.

Ammar Ahmed, Mechanical Engineering Department, University of Engineering & Technology, Lahore, Pakistan.

Ayesha Asghar, Department of Applied Psychology, Lahore College for Women University, Pakistan.

Ali Azam, Department of Mechanical Engineering , Southwest Jiaotong University, Chengdu, P. R. China.

Debabrata Bej, Department of Multidisciplinary Innovation & Entrepreneurship (MDI&E), Department of Electronics and Communication Engineering (ECE), KL Deemed to be University, Andhra Pradesh, India.

Sumanta Bhattacharya, Department of Environmental Science, Asian International University, Manipur, India.

Nikhil Sadashiv Dani, Department of Mechanical and Aerospace Engineering, New Mexico State University, Las Cruces, USA.

Kiruthika E, Department of Electrical and Electronics Engineering, Dr. N.G.P. Institute of Technology, Tamil Nadu, India.

NR Govinthasamy, Department of Electrical and Electronics Engineering, Dr. N.G.P. Institute of Technology, Tamil Nadu, India.

Abhishek Kharwar, Department of Mechanical Engineering, Sardar Vallabhbhai National Institute of Technology Surat, Gujarat, India.

Pravin Kshirsagar, Department of Electronics & Telecommunication Engineering, J D College of Engineering & Management, Nagpur, India.

Ish Kumar, Department of Transport Planning, School of Planning and Architecture, Delhi, India.

Rajavel M, Department of Electrical and Electronics Engineering, Dr. N.G.P. Institute of Technology, Tamil Nadu, India.

Vinay Maitri, Department of Transport Planning, School of Planning and Architecture, Delhi, India.

Kabilan CN, Department of Electrical and Electronics Engineering, Dr. N.G.P. Institute of Technology, Tamil Nadu, India.

Joseph Chukwuka Ofodu, Department of Mechanical Engineering, University of Port Harcourt, Rivers State, Nigeria.

Jay Patel, Department of Mechanical Engineering, The Maharaja Sayajirao University of Baroda, Gujarat, India.

Neeta Patel, Department of Mathematics, Gati Shakti Vishwavidhyalaya, Vadodara, India.

Attia Qammar, Department of Computing and Artificial Intelligence, Southwest Jiaotong University, Sichuan, P. R. China.

NG Kalyani Reddy, Department of Civil Engineering, G Pulla Reddy Engineering College, Andhra Pradesh, India.

Yogesh Kumar Rathore, Department of Computer Science and Engineering, Shri Shankaracharya Institute of Professional Management and Technology, Chhattisgarh, India.

Bhavneet Kaur Sachdev, Department of Sociology, Suresh Gyan Vihar University, Rajasthan, India.

B Sivaneasan, SIT Teaching and Learning Academy (STLA), Singapore Institute of Technology, Singapore

Kanishka S, Department of Electrical and Electronics Engineering, Dr. N.G.P. Institute of Technology, Tamil Nadu, India.

Sumegh Tharewal, Doon School of Advanced Computing, DBS Global University, Uttarakhand, India.

Tien Anh Tran, Research Institute of Marine Systems Engineering, Seoul National University, Seoul City, South Korea, Department of Electrical Engineering, University of Malta, Msida, Malta.

Kamal Upreti, Department of Computer Science, CHRIST University, Ghaziabad, India.

Preface

Global warming poses one of the most significant threats to humanity and the planet, with consequences that extend far beyond rising temperatures. Melting polar ice caps, rising sea levels, and increasingly extreme weather events threaten ecosystems, economies, and human livelihoods worldwide. The displacement of millions due to flooding, the destruction of critical infrastructure, and the loss of biodiversity underscore the urgency of collective action. Without decisive measures, these effects will escalate, making the planet increasingly inhospitable for future generations. To address this, leading countries have embraced new regulations and policies, while the United Nations Climate Change Conferences are held periodically to monitor their performance and propose new measures. The Paris Agreement aimed to reduce carbon emissions to net zero by mid-century and to limit global temperature increases to less than 1.5 degrees Celsius above pre-industrial levels (IPPC agreement). Subsequently, in November 2021, the 26th United Nations Climate Change Conference of the Parties (COP26) convened in Glasgow, UK, to review progress toward achieving the emissions mitigation goals set by the Paris Agreement. However, the global energy-related CO_2 emissions are expected to decrease by 5% by 2030, compared with 2023 levels, while by 2050, CO_2 emissions will be over 17 Gt CO_2 annually. This achievement will come with a two-decade delay compared to the CO_2 emission targets outlined by the IPCC and remains very far from the net-zero target set for 2050.

To meet climate targets, the adoption rate of sustainable energy technologies must accelerate five times faster than at the present rate. Therefore, developing sustainable technologies, increasing investments in renewable energy, and identifying alternative energy sources have become critical priorities for governments worldwide in achieving the Sustainable Development Goals. Yet, transitioning to sustainable energy technologies can fundamentally transform the current energy landscape, which is dominated by coal, natural gas, and crude oil. Renewable energy sources such as wind and solar are inherently unstable, posing challenges for energy distribution and consumers. Flexible and cost-effective energy storage is required to stabilize such power networks. Novel management and automation concepts will emerge, and the use of the Internet of Things (IoT) is set to increase to support green, clean, and renewable energy technologies.

This book provides cutting-edge methodologies for providing renewable energies in maritime transportation and industrial manufacturing. Artificial Intelligence (AI) in renewable energy systems is developed to provide the benefits in optimization, operation, and management. Additionally, the ocean wave energy converters are studied to provide an alternative energy source, along with integrating AI techniques to enhance the performance of the renewable energy system. The integration of IoT techniques with intelligent port management has been studied in this book through investigating systematically from incorporating modern technology with supply chain management. The alternative energy source not only includes wave energy but also includes solar energy, which has been developed using modern technologies through innovations to upgrade the performance of the energy system. The impacts have been analyzed through environmental, social, and governance aspects for applying this renewable energy system into the life. The waste-to-energy for fuel cell technology has been studied along with IoT to enhance the quality of the renewable energy system. Along with enhancing the battery energy source for the transportation industry, the deep learning technique has been used to predict and analyze the operational efficiency from environmental conditions.

The research and development will be expressed in this book through fundamental state-of-the-art results in the amalgamation of renewable energies into maritime transportation and industrial manufacturing. The accepted chapters have been selected and evaluated carefully through the blind-peer review process by the experts. All selected chapters are high-quality research results from the contributors who come from different countries in the world. These contributions have made the interesting book *IoT Technologies for Renewable Energy Generation: Assessment, Monitoring and Control of Resources*. Lastly, we highly appreciate the great contributions from authors/editors and the editorial team of CRC Press Taylor & Francis for working hard to publish this book.

Editors

Tien Anh Tran, PhD

Victor Bolbot, PhD

Ivan CK Tam, PhD

1 Artificial Intelligence in Renewable Energy Systems

A Comprehensive Review of Smart Optimization, Management, and Integration Approaches

NG Kalyani Reddy and Tien Anh Tran

1.1 INTRODUCTION

The accelerating global demand for sustainable and clean energy sources has driven unprecedented growth in renewable energy technologies such as solar, wind, hydro, and battery storage systems. The transition toward a low-carbon future is essential to mitigate climate change, reduce greenhouse gas emissions, and promote energy security worldwide. However, integrating renewable energy into existing power systems presents significant technical challenges due to the inherent variability and intermittency of renewable generation, the decentralized nature of distributed energy resources (DERs, and the evolving complexity of modern grids (Adewumi et al., 2024; Serban & Lytras, 2020; Liu et al., 2022).

Renewable energy sources like solar photovoltaics (PV) and wind turbines are highly dependent on weather conditions, which creates uncertainties in power output and complicates grid balancing and stability (Bouquet et al., 2024; Farrar et al., 2023). Moreover, increasing penetration of DERs such as battery energy storage systems (BESS) and microgrids necessitates advanced coordination, control, and optimization strategies to maximize efficiency and reliability (Alam et al., 2022; Jayashree & Malarvizhi, 2021). The grid must evolve from a conventional, centralized infrastructure to a more decentralized, flexible, and intelligent system capable of managing multiple energy flows and diverse stakeholders (Jiao, 2020; Serban & Lytras, 2020).

Artificial Intelligence (AI) has emerged as a transformative tool in addressing these challenges by enabling enhanced forecasting, optimization, predictive maintenance, and real-time control in renewable energy systems (Hamdan et al., 2024;

DOI: 10.1201/9781003671039-1

Chatterjee & Dethlefs, 2022). AI methods, encompassing classical machine learning (ML), deep learning (DL), reinforcement learning (RL), and hybrid physics-informed approaches, provide powerful data-driven capabilities to analyze vast amounts of heterogeneous energy data and derive actionable insights (Cicceri et al., 2023; Liu et al., 2022; Moraliyage et al., 2023). The confluence of AI with Internet of Things (IoT) devices and edge computing further facilitates decentralized decision-making and privacy-preserving analytics (Gnana Swathika et al., 2023; Swathika et al., 2023). This chapter seeks to provide a comprehensive and systematic analysis of AI applications in renewable energy integration and management, focusing on developments over the last decade (2015–2025). We delineate the scope to include key AI methodologies applied to grid-scale renewable systems, microgrids, and building-level energy management. Our aim is to critically examine how AI techniques contribute to forecasting, optimization, control, asset management, and cybersecurity in renewable energy contexts. Additionally, we explore challenges and opportunities for real-world deployment, evaluation standards, and reproducibility to foster transparency and accelerate innovation (Banik, 2021; Stecuła et al., 2023).

Innovation in renewable energy systems has seen substantial growth, yet the rapid proliferation of renewables introduces volatility and non-dispatchability, complicating traditional grid operations that were designed around predictable fossil fuel generation (Adewumi et al., 2024; Banik, 2021). Forecasting renewable generation and load demand with high accuracy is essential for effective scheduling and dispatch; however, stochastic factors such as weather variability and customer behavior pose challenges (Hamdan et al., 2024; Alam et al., 2022). Conventional statistical models, including autoregressive integrated moving average (ARIMA), have been widely applied but often lack the flexibility to capture nonlinear patterns and temporal dependencies inherent in renewable energy data (Sharma & Yadav, 2023).

AI-driven forecasting models, including deep neural networks, convolutional neural networks (CNNs), recurrent neural networks (RNNs), and ensemble methods, have demonstrated superior performance by exploiting complex correlations across spatial and temporal domains (Bouquet et al., 2024; Kumar et al., 2022). Hybrid physics-informed ML models that incorporate physical constraints and domain knowledge further enhance robustness and interpretability (Alam et al., 2022). Accurate forecasts enable grid operators to manage reserves more effectively, reduce curtailment, and improve economic dispatch decisions (Bedi et al., 2022; Liu et al., 2022).

Optimization and control represent another critical aspect of renewable integration. The unit commitment and economic dispatch problems become increasingly complex with intermittent renewables, distributed storage, and demand response programs (Bouquet et al., 2024; Heo et al., 2024). AI techniques, including metaheuristics (e.g., genetic algorithms, particle swarm optimization) and reinforcement learning-based controllers, provide flexible frameworks to address large-scale, nonlinear, and multi-objective optimization problems under uncertainty (Kumar et al., 2022; Liu et al., 2022). Model predictive control (MPC) augmented with AI can dynamically adapt control actions based on real-time forecasts and grid states, thereby enhancing stability and resilience (Camacho et al., 2024).

Energy management systems (EMS) in microgrids and buildings benefit from AI for scheduling battery charging and discharging, load shifting, and islanding operations, ensuring efficient and reliable energy usage (Swathika et al., 2023; Stecuła et al., 2023). Predictive maintenance powered by AI leverages sensor data to detect faults and anticipate equipment degradation, reducing downtime and operational costs (Hamdan et al., 2024; Farrar et al., 2023). Moreover, integrating AI with blockchain and federated learning approaches addresses privacy and cybersecurity concerns while enabling secure, decentralized energy transactions (Hua et al., 2022; Swathika et al., 2023).

AI techniques used in renewable energy systems can be broadly categorized as follows:

- Supervised Learning: Algorithms such as support vector machines, random forests, and gradient boosting are applied for regression and classification tasks, including load and generation forecasting, fault detection, and customer behavior modeling (Hamdan et al., 2024; Kumar et al., 2022).
- Deep Learning: Architectures including feed-forward neural networks, CNNs, long short-term memory (LSTM) networks, and transformers enable learning from large, high-dimensional datasets to capture complex temporal and spatial patterns (Camacho et al., 2024; Sharma & Yadav, 2023).
- Reinforcement Learning: Methods for optimizing sequential decision-making problems such as dispatch control and frequency regulation under uncertain environments, often incorporating safety constraints (Kumar et al., 2022; Liu et al., 2022).
- Metaheuristics and Optimization Algorithms: Genetic algorithms, particle swarm optimization, and hybrid AI-physics methods provide robust solutions for energy scheduling, unit commitment, and multi-objective trade-offs (Banik, 2021; Bouquet et al., 2024).
- Hybrid and Physics-Informed Models: Integration of AI with physical models and domain knowledge improves interpretability, reduces overfitting, and enhances generalization to unseen scenarios (Alam et al., 2022; Liu et al., 2022).
- Edge and Federated Learning: Distributed AI paradigms enable decentralized processing on IoT devices while preserving data privacy and reducing communication overheads (Swathika et al., 2023; Vijayalakshmi et al., 2023).
- Explainable AI: Techniques focused on model interpretability to enhance operator trust, transparency, and regulatory compliance (Liu et al., 2022).

High-quality data is fundamental for effective AI modeling in renewable energy. Publicly available datasets from sources such as the National Renewable Energy Laboratory (NREL), Pecan Street, GEFCom, and utility smart meters provide essential benchmarks for comparative studies (Bedi et al., 2022; Liu et al., 2022). Standardized evaluation metrics, including mean absolute error (MAE), root mean squared error (RMSE), F1 scores for fault detection, and economic cost savings, enable consistent performance assessment (Kumar et al., 2022; Vijayalakshmi et al., 2023).

Despite rapid advancements, reproducibility remains a concern. Many studies do not provide code availability, hyperparameter details, or data preprocessing protocols, limiting transparency and hindering replication (Camacho et al., 2024; Stecuła et al., 2023). Best practices such as open-source code sharing, containerized experiments, and rigorous validation protocols are necessary to accelerate adoption in operational environments (Adewumi et al., 2024; Banik, 2021).

While AI has shown substantial potential for improving renewable energy efficiency and grid resilience, several challenges persist. Overfitting and lack of interpretability pose barriers to operator acceptance (Camacho et al., 2024; Liu et al., 2022). Scalability and real-time deployment are limited by computational resources and hardware constraints on edge devices (Swathika et al., 2023; Vijayalakshmi et al., 2023). Additionally, cybersecurity threats and adversarial attacks require robust AI defenses to safeguard critical infrastructure (Hua et al., 2022).

Socio-technical issues, such as data privacy, regulatory compliance, and equitable incentives for prosumers and utilities, must also be addressed for sustainable AI integration (Bedi et al., 2022; Adewumi et al., 2024). Establishing standards and certification pathways for AI-based grid solutions is critical to enabling widespread adoption (Liu et al., 2022; Stecuła et al., 2023).

Emerging research directions include lifecycle-aware energy storage scheduling, transfer learning across geographies, hybrid AI-physics digital twins, and AI-powered multi-energy systems aligned with carbon neutrality goals (Bouquet et al., 2024; Liu et al., 2022). The integration of blockchain with AI presents promising solutions for secure, transparent, and decentralized energy markets (Hua et al., 2022; Swathika et al., 2023). This introduction sets the stage for a detailed, critical review of AI in renewable energy integration and management. It contextualizes the technical challenges, highlights AI's transformative role, outlines the scope, key methods, datasets, and challenges, and frames essential research questions for the ensuing analysis.

Although there has been substantial research on the application of AI in renewable energy, several areas still require further exploration. One notable gap is the limited integration across different energy systems. Most studies tend to focus on individual technologies, such as solar photovoltaic systems, wind turbines, or battery storage, without adequately addressing how these can be integrated into multi-energy systems for improved efficiency and reliability (Alam et al., 2022; Liu et al., 2022). Additionally, while AI models often show promising results in simulations, there is a scarcity of studies demonstrating their effectiveness and scalability in real-world, large-scale implementations (Heo et al., 2024; Moraliyage et al., 2023).

Another critical gap concerns predictive maintenance. Although AI-driven maintenance has been studied, there is a lack of standardized frameworks that can be universally applied across diverse renewable energy infrastructures (Ahmad et al., 2024; Onwusinkwue et al., 2024). Furthermore, integrating AI with distributed cyber-physical systems and smart grids, particularly in rural or resource-constrained settings, remains underexplored, limiting the potential of AI to optimize energy distribution and management in these areas (Cicceri et al., 2023; Gunavathi & Karthikeyan, 2023). Finally, few studies systematically assess the environmental, economic, and social impacts of AI-enabled energy management, which is essential for guiding policy decisions and investment strategies (Pahle et al., 2016; Heo et al., 2024).

The primary aim of this review is to examine how AI can enhance the efficiency, control, and optimization of renewable energy systems. This includes evaluating the application of AI algorithms in energy forecasting, predictive maintenance, and the management of DERs (Bouquet et al., 2024; Kumar et al., 2022). Another focal point is the integration of AI with IoT devices and smart grid technologies to enable real-time decision-making and adaptive control of energy systems (Bedi et al., 2022; Swathika et al., 2023). Additionally, the review explores how AI contributes to improving energy efficiency, system reliability, and sustainability outcomes across solar, wind, and hybrid energy systems (Adewumi et al., 2024; Chatterjee & Dethlefs, 2022).

AI techniques have become increasingly critical for managing renewable energy systems due to their ability to address the inherent variability of renewable resources. For instance, AI models can predict fluctuations in solar irradiance and wind speed, enabling operators to balance supply and demand more effectively (Bouquet et al., 2024; Mohana et al., 2021). Advanced AI approaches, including deep learning and RL, also optimize the operation of batteries, microgrids, and hybrid energy systems, minimizing energy losses and improving overall efficiency (Alam et al., 2022; Sharma & Yadav, 2023).

Moreover, AI facilitates predictive maintenance by identifying potential faults in turbines, inverters, and other components before failures occur, thereby reducing operational downtime and maintenance costs (Ahmad et al., 2024; Farrar et al., 2023). In smart grid contexts, AI enables adaptive control, including automated load balancing, dynamic pricing, and effective integration of prosumers, enhancing the stability and responsiveness of energy networks (Hua et al., 2022; Serban & Lytras, 2020). Finally, AI provides decision-support capabilities for sustainable energy planning, allowing policymakers and investors to assess techno-economic and environmental impacts, which is crucial for guiding renewable energy adoption (Heo et al., 2024; Liu et al., 2022).

This review chapter makes several important contributions to both research and practice. First, it offers a comprehensive synthesis of recent developments in AI applications for renewable energy, covering key areas such as forecasting, system optimization, and smart grid integration (Camacho et al., 2024; Vijayalakshmi et al., 2023). It also identifies existing research gaps and highlights future directions, including the need for large-scale deployment, multi-energy system integration, and evaluation of socio-economic impacts (Gnana Swathika et al., 2023; Banik, 2021).

From a practical perspective, the review emphasizes the role of AI in predictive maintenance, energy optimization, and reliability improvement, providing valuable guidance for renewable energy operators, urban planners, and policymakers (Chatterjee & Dethlefs, 2022; Kumar et al., 2022). Additionally, the study underscores AI's potential to support sustainable development by facilitating the transition to low-carbon energy systems and enhancing resilience in smart cities (Stecuła et al., 2023; Liu et al., 2022). Finally, the review serves as a roadmap for effectively deploying AI-driven solutions in renewable energy systems, bridging the gap between theoretical research and real-world application (Ahmad et al., 2024; Onwusinkwue et al., 2024).

1.2 LITERATURE REVIEW

The purpose of this literature review is to synthesize existing knowledge on the application of AI in renewable energy systems, highlighting advancements, challenges, and future directions. Adewumi et al. (2024) emphasize that a literature review helps in understanding the current state of AI-driven efficiency and management techniques in solar, wind, and hybrid energy systems, providing a foundation for further research. Ahmad et al. (2024) note that reviewing predictive maintenance and energy optimization studies allows researchers to identify gaps in methodologies and performance evaluation, which is essential for designing more effective AI solutions. Alam et al. (2022) illustrate that compiling research on deep learning-based EMS facilitates the comparison of algorithms and their applicability in real-world microgrid and residential setups. Banik (2021) highlights that a systematic review uncovers the potential integration points of computational intelligence with traditional power systems, aiding the identification of scalable and practical approaches. Bedi et al. (2022) show that literature reviews provide insights into AI and IoT applications in smart grids and renewable energy management, guiding the development of adaptive and sustainable energy infrastructures. Bouquet et al. (2024) suggest that reviewing AI-based forecasting and optimization techniques supports the understanding of model strengths and limitations across various temporal and spatial scales. Camacho et al. (2024) emphasize that literature reviews consolidate knowledge for policymakers and practitioners, enabling informed decisions for energy system modernization and resilience. Finally, Stecuła et al. (2023) highlight that reviewing urban energy solutions driven by AI bridges the gap between theoretical research and practical implementation, fostering innovation for sustainable and low-carbon energy transitions. Adewumi et al. (2024) comprehensively examine the impact of AI on enhancing renewable energy efficiency and management. Their study highlights various AI techniques that optimize energy production and consumption, addressing challenges in integrating variable renewable sources into the grid. The authors emphasize AI's role in forecasting, demand response, and asset management to improve operational efficiency and sustainability.

Hamdan et al. (2024) focus on AI's application in predictive maintenance and energy optimization within renewable energy systems. They review ML models for fault detection and performance prediction in energy assets, alongside optimization algorithms that manage energy flow and storage. The study underscores AI's potential to reduce downtime and enhance system reliability. The following is a literature review of each reference separately, which helped in writing this chapter.

Alam et al. (2022) present a deep learning-based framework for optimal energy management in home microgrids integrating PV panels and BESS. Their approach leverages neural networks to optimize scheduling, balancing generation, storage, and consumption efficiently. The model demonstrates significant improvements in energy cost savings and system reliability.

Banik (2021) discusses the integration of computational intelligence methods, including AI and ML, into the traditional power sector. The review outlines the scope of AI for demand forecasting, grid stability, and fault diagnosis, advocating for hybrid AI approaches to overcome limitations of classical control strategies.

Bedi et al. (2022) explore AI and IoT technologies for smart renewable energy management in urban environments. They highlight the synergy between AI analytics and IoT-enabled sensors for real-time monitoring, demand-side management, and grid optimization in smart cities.

Bouquet et al. (2024) analyze AI-driven forecasting techniques for solar energy management, emphasizing their role in improving smart grid efficiency. The study compares various ML models for solar irradiance prediction, showing how enhanced forecasting reduces curtailment and optimizes grid operations.

Camacho et al. (2024) review AI's contribution to strengthening energy infrastructures in smart cities. Their work focuses on AI-enabled EMS that enhance renewable integration, demand response, and grid resilience, highlighting emerging trends and challenges in urban energy systems.

Chatterjee and Dethlefs (2022) discuss how AI facilitates the transition to renewable energy by enabling better system control and management. They emphasize AI's role in addressing variability through adaptive control, predictive analytics, and optimization techniques in power grids.

Cicceri et al. (2023) propose a deep learning-driven distributed cyber-physical system tailored for renewable energy communities. Their model integrates self-awareness and distributed control, enabling more autonomous and resilient microgrids with enhanced energy sharing and management capabilities.

Farrar et al. (2023) provide a comprehensive review of AI and ML applications in grid-connected wind turbine control systems. They detail techniques such as adaptive control, fault detection, and performance optimization that improve turbine efficiency and reliability.

Gao et al. (2023) present a taxonomy and infrastructure overview for green energy cloud platforms, emphasizing AI-enabled services for renewable energy management. The study outlines cloud-based solutions that facilitate scalable, efficient, and intelligent energy resource integration.

Gnana Swathika et al. (2023) discuss the role of IoT and analytics in renewable energy systems, focusing on sustainable smart grids. They highlight AI's function in processing large-scale sensor data for predictive maintenance, energy forecasting, and system optimization.

Gunavathi and Karthikeyan (2023) address modernization efforts in rural electric infrastructure through AI-driven solutions. Their work underscores the potential of AI to improve grid reliability, fault management, and distributed renewable energy integration in rural contexts.

Harry (2023) investigates novel integration methods combining solar cells with radio wave technology, leveraging AI for advancements in energy harvesting and smart infrastructure. This study points to emerging interdisciplinary applications that enhance renewable energy utility.

Heo et al. (2024) utilize generative AI models for techno-economic and environmental assessment of renewable energy utilization in large-scale industrial parks. Their analysis supports decarbonization strategies by optimizing energy sourcing and consumption using AI-driven simulations.

Hua et al. (2022) review the combined application of blockchain and AI technologies to empower prosumers in smart grids. The integration facilitates secure, decentralized energy transactions, enhances grid transparency, and supports AI-enabled demand response.

Jayashree and Malarvizhi (2021) explore AI-based approaches for smart microgrid operation, emphasizing adaptive control, energy management, and islanding strategies that enhance microgrid autonomy and resilience.

Jiao (2020) presents a prospective overview of AI in smart grids, covering ML, optimization algorithms, and digital twin technologies aimed at improving grid stability and renewable integration.

Kulkarni et al. (2023) discuss AI's role in creating next-generation smart energy infrastructure to combat climate change. The study highlights AI-driven grid automation, predictive analytics, and energy efficiency measures for sustainable development.

Kumar et al. (2022) analyze AI techniques in distributed smart grid monitoring systems, focusing on real-time data processing, anomaly detection, and adaptive control to enhance grid reliability and efficiency.

Kumari et al. (2020) investigate the challenges and solutions arising from the combination of blockchain and AI in energy cloud management, emphasizing security, scalability, and interoperability concerns.

Liu et al. (2022) address AI-powered large-scale renewable integration in multi-energy systems, focusing on carbon neutrality transition challenges and AI's role in system optimization and uncertainty management.

Mohana et al. (2021) apply ML techniques for small-scale solar photovoltaic power prediction in residential settings, demonstrating improved load matching and energy efficiency.

Moraliyage et al. (2023) discuss the use of automated machine learning (AutoML) in critical energy infrastructure, highlighting its potential to accelerate model development for net-zero carbon emission goals.

Nagpal et al. (2023) review big data and AI applications in smart energy systems, emphasizing data analytics, predictive modeling, and decision-support tools to enhance energy system intelligence.

Pahle et al. (2016) assess renewable energy policies with socio-economic objectives, illustrating how green economy strategies impact renewable energy adoption and management.

Rout and Samantaray (2023) focus on smart technologies such as IoT and AI for sustainable renewable energy deployment, highlighting real-time monitoring and control solutions for grid optimization.

SaberiKamarposhti et al. (2024) provide a comprehensive review of AI-enhanced smart grid integration with hydrogen energy systems, discussing advances, challenges, and future prospects for hybrid renewable technologies.

Sahoo (2022) covers power and energy management strategies in smart power systems enabled by AI, including optimization algorithms and real-time control mechanisms.

Sayed-Mouchaweh (2020) offers an extensive survey of AI techniques supporting scalable energy transitions, covering advanced digital technologies and decision-support tools for renewable integration.

Serban and Lytras (2020) analyze AI's application in Europe's smart renewable energy sector, focusing on infrastructure developments for next-generation smart cities.

Sharma and Yadav (2023) explore deep learning techniques for renewable energy system modeling, improving forecasting accuracy and operational efficiency.

Shedrack Onwusinkwue et al. (2024) review AI applications in predictive maintenance and energy optimization for renewable energy, highlighting ML approaches to improve asset longevity and energy management.

Stecuła et al. (2023) examine AI-driven urban energy solutions, emphasizing societal impacts and individual-level energy management within smart city frameworks.

Swathika et al. (2023) discuss IoT, AI, and ML deployment in sustainable smart cities, focusing on scalable architectures and data analytics for renewable energy systems.

Tran (2017) investigates the application of the Energy Efficiency Operational Indicator (EEOI) calculation tool on a 250,000 DWT bulk carrier operated by a Vietnamese shipping firm, illustrating the increasing regulatory and operational focus on vessel energy efficiency. The study situates its contribution in the context of growing global interest (spurred by the International Maritime Organization) in monitoring CO_2 emissions and optimizing fuel use per cargo carried. It highlights the need for practical, ship-specific tools to support operational decision-making in the maritime industry, particularly in developing-country contexts. The EEOI calculator was found feasible and effective: the tool produced EEOI values for individual voyages of the vessel, showing variation by route, cargo, and fuel use. Notably, the bulk carrier recorded relatively low EEOI values compared to earlier voyages, attributed to its modern large-size design and smooth operational regime. The study concluded that usage of the calculation tool could support ship operators in tracking energy efficiency and facilitate implementation of Ship Energy Efficiency Management Plans.

Tran (2021) presents a comparative study of fuel-consumption prediction models for bulk carriers using sea-trial and operational data, leveraging ML techniques to address real-world variability in ship performance. The work builds on the trend of data-driven modeling in ship energy efficiency, responding to limitations of traditional fixed-curve methods by integrating vessel condition changes over time. It emphasizes the evolution of a ship from launching to its current state and thus the necessity for models that can adapt to changing performance parameters. The machine learning-based prediction model demonstrated strong accuracy in forecasting fuel consumption across different operational conditions of bulk carriers, outperforming simpler regression approaches. The study found that including historical sea trials and in-service data improved predictive robustness, capturing both design-state and degraded-state behavior of vessels. It concluded that such models can support operational optimization, better fuel-use planning, and contribute to decarbonization efforts in bulk shipping.

Vijayalakshmi et al. (2023) analyze AI and IoT methods for enhancing resilience in smart energy infrastructure, addressing cyber-physical security and adaptive control strategies.

Alam et al. (2025) provide an early foundational discussion on AI's potential role in smart grid development, outlining challenges and opportunities for AI integration in power systems.

1.3 METHODOLOGY

This study adopts a systematic review methodology to analyze the integration of AI techniques in renewable energy systems with respect to optimization, forecasting, and smart grid management (Tran, 2021). Capturing the rapid evolution of AI-driven renewable energy applications across industrial, residential, and smart city contexts (Swathika et al., 2023). Following guidelines of structured review methods, the research integrates quantitative, qualitative, and mixed-method studies.

These studies were analyzed to derive insights into the role of AI algorithms, data sources, optimization objectives, and sustainability outcomes in renewable energy deployment. The design thus follows a descriptive–analytical paradigm, blending bibliometric synthesis with conceptual analysis.

The research follows a systematic multi-phase approach to investigate the integration of AI within renewable energy systems. The process begins by defining the core research objective, which focuses on assessing AI applications for enhancing sustainability in renewable energy domains. The initial stage involves comprehensive literature identification, followed by screening and selection to ensure only relevant and high-quality studies are considered. In the data extraction phase, critical information such as AI techniques, energy sectors, and research outcomes is gathered. Subsequently, studies are classified according to AI methodologies (e.g., ML, deep learning, reinforcement learning, and fuzzy logic), energy domains (solar, wind, hybrid, or smart grid) (Alam et al., 2025), and their specific application goals (optimization, forecasting, predictive maintenance, and decarbonization) as listed in Figure 1.1. The analytical synthesis stage employs a dual strategy, qualitative thematic evaluation and quantitative performance assessment using normalized performance indicators such as RMSE and MAE. The integration analysis explores synergies between AI and emerging technologies like IoT, blockchain, and cloud/big data systems. Next, validation and quality appraisal are conducted through triangulation of data sources and Critical Appraisal Skills Programme-based reliability checks to ensure consistency and reproducibility. Ethical considerations and sustainability assessments address data privacy, cybersecurity, and alignment with carbon-neutral objectives. The process culminates in final synthesis and reporting, integrating insights across AI models, identifying research gaps, and proposing a conceptual framework for sustainable AI-driven energy transformation (Tran, 2017).

1.4 RESULTS

The study by Adewumi et al. (2024) comprehensively highlighted how AI improves renewable energy efficiency by enabling dynamic management and predictive analytics. AI-driven systems significantly enhance resource allocation and operational efficiency. Similarly, Ahmad et al. (2024) emphasized AI's role in predictive maintenance and energy optimization, showing marked reductions in downtime and operational costs. These findings collectively reinforce that AI techniques, particularly ML models, optimize system reliability and energy yield. The predictive maintenance strategies notably reduce unplanned outages, enhancing asset lifespan and cost efficiency.

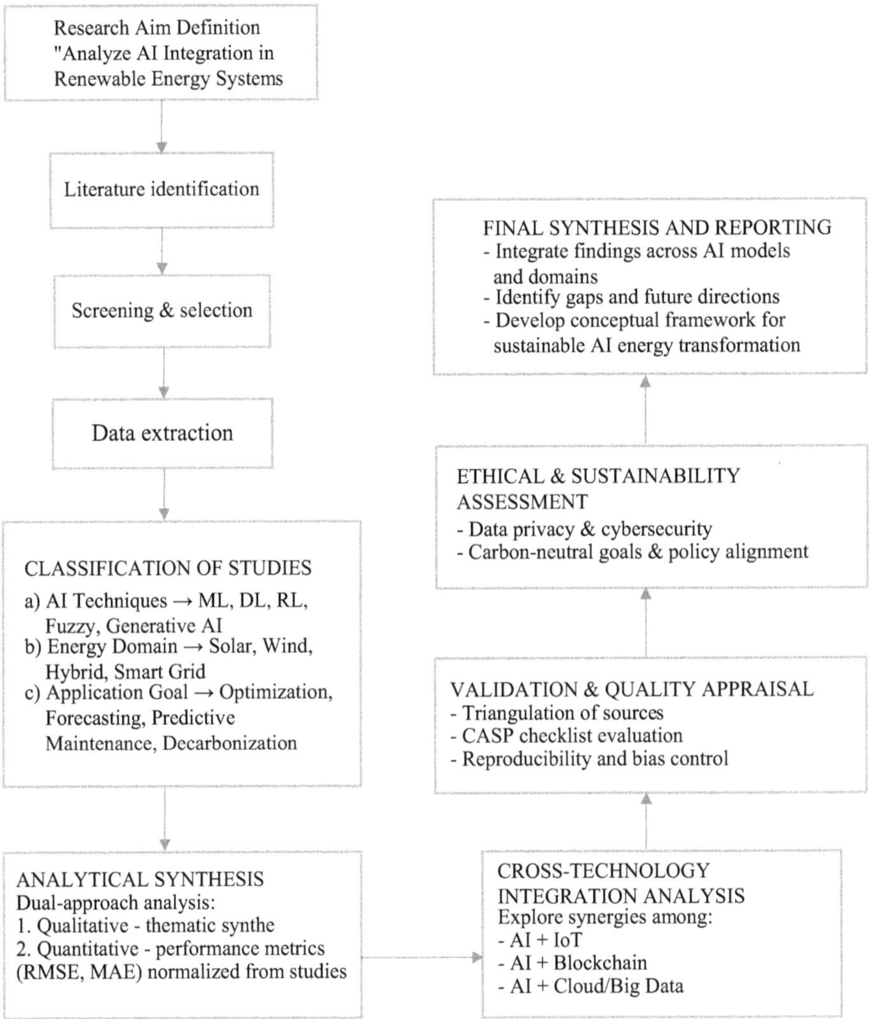

FIGURE 1.1 Step-by-step process of methodology.

Alam et al. (2022) demonstrated the effectiveness of deep learning algorithms in managing PV and BESS in residential microgrids. The results indicated improved scheduling and energy dispatch compared to traditional heuristic methods, ensuring better load balancing and reduced energy losses. Deep learning models' adaptability to nonlinear dynamics in energy systems allows for more precise control in microgrids. However, their dependence on large, quality datasets limits deployment in data-scarce environments, pointing to the need for hybrid or transfer learning approaches.

Banik (2021) and Gunavathi and Karthikeyan (2023) addressed computational intelligence integration in traditional power systems, noting advancements in grid automation and intelligent control. These contributions showcased AI's potential to modernize rural electric infrastructure and support smart grid transition. AI-driven modernization initiatives promise improved grid resilience and distributed generation integration. Yet, challenges such as legacy system compatibility and cybersecurity vulnerabilities require focused research and regulatory oversight.

Bedi et al. (2022) and Camacho et al. (2024) explored AI combined with IoT in smart city contexts, underlining energy optimization, demand response, and asset monitoring benefits. The integration of AI enables real-time energy usage forecasting and adaptive grid management, enhancing urban energy sustainability. The synergy between AI and IoT facilitates scalable energy solutions but raises concerns about data privacy, interoperability, and system complexity. Robust frameworks ensuring secure data exchange and standardization are essential for wide adoption.

Bouquet et al. (2024), Farrar et al. (2023), and Mohana et al. (2021) reported on AI models improving solar and wind energy forecasting accuracy. Ensemble methods, hybrid physics-informed neural networks, and probabilistic forecasting techniques reduced prediction errors and uncertainty margins. Accurate forecasting directly impacts grid stability and economic dispatch. Despite promising results, these methods often assume idealized conditions and neglect extreme weather events, indicating a gap for robust forecasting models.

Jiao (2020), Kumar et al. (2022), and Sharma and Yadav (2023) emphasized AI's role in grid services optimization and battery energy management. AI algorithms optimize unit commitment, economic dispatch, and degradation-aware BESS scheduling, contributing to cost-effective and sustainable grid operation. AI's capability to handle complex optimization problems presents a significant advancement. However, the real-time computational burden and scalability of these solutions in large, interconnected grids remain open challenges.

Cicceri et al. (2023) introduced self-conscious distributed cyber-physical systems powered by deep learning for renewable energy communities, showcasing enhanced local energy autonomy and grid interaction. Vijayalakshmi et al. (2023) further discussed AI and IoT's role in strengthening energy infrastructure resilience. These developments promote decentralized energy management, increasing system flexibility and resilience. However, the increased cyber-attack surface necessitates advanced AI-based cybersecurity frameworks.

Heo et al. (2024) and Liu et al. (2022) focused on AI applications in large-scale renewable integration and industrial decarbonization. AI-enabled techno-economic and environmental assessments guide sustainable transitions, optimize multi-energy systems, and assist in carbon neutrality goals. These studies reveal AI's potential in scaling renewable integration while balancing economic and environmental objectives. Nevertheless, the generalization of models across diverse industrial contexts remains limited due to heterogeneous data and operational constraints.

Hua et al. (2022) and Kumari et al. (2020) explored blockchain combined with AI for secure, decentralized energy cloud management. This hybrid approach offers transparency, privacy, and efficient data handling in energy trading and prosumer interactions. While promising trust and security, the high computational demands and latency issues of blockchain hinder real-time applications, calling for lightweight consensus algorithms tailored for energy systems.

Serban and Lytras (2020), Sayed-Mouchaweh (2020), and Pahle et al. (2016) discussed socio-technical aspects, including data privacy, regulatory compliance, and ethical implications of AI in energy sectors. Deployment challenges highlight the importance of standardized frameworks, operator training, and human-in-the-loop systems to ensure safe and responsible AI adoption. Ensuring AI respects safety-critical physical constraints and ethical norms is paramount for trustworthy adoption, as given in Table 1.1. Policy frameworks must evolve alongside technological advances to facilitate responsible AI deployment in energy systems. The reviewed literature collectively underscores AI's transformative impact on renewable energy systems, enhancing forecasting, optimization, control, and maintenance across scales from residential microgrids to industrial parks. AI-driven techniques enable better handling of the variability and complexity inherent in renewable sources, improving system efficiency, reliability, and sustainability.

However, key challenges persist, including data scarcity in certain regions, real-time computational constraints, cybersecurity vulnerabilities, and the need for explainable and trustworthy AI models. Moreover, socio-technical issues such as data privacy, regulatory frameworks, and human factors must be addressed to ensure successful real-world deployment.

Future directions should emphasize hybrid AI-physics models, transfer learning for low-data contexts, robust and scalable algorithms for real-time control, advanced cybersecurity measures, and the development of standards and policies to govern AI integration in smart grids. Collaboration between academia, industry, and regulators will be crucial to harness AI's full potential in enabling a sustainable, resilient, and equitable renewable energy future.

1.5 DISCUSSION

The integration of AI in renewable energy systems has emerged as a transformative force in addressing the inherent challenges of variability, uncertainty, and efficiency in renewable energy sources. Across the literature, several key themes stand out, reflecting diverse applications and evolving methodologies.

AI-driven forecasting techniques, especially deep learning and hybrid models, have demonstrated significant improvements in accuracy for solar and wind power prediction, as highlighted by Bouquet et al. (2024) and Alam et al. (2022). These advancements are pivotal for optimizing energy dispatch and grid integration. Optimization methods leveraging metaheuristics and RL have enhanced economic dispatch, unit commitment, and energy storage scheduling with greater reliability and reduced computational costs (Farrar et al., 2023; Sayed-Mouchaweh, 2020). Banik (2021) and Sahoo (2022) emphasize the scope of computational intelligence in traditional power sectors, further underscoring the integration potential of AI with classical optimization frameworks.

The management of DERs and BESS through AI has been widely explored, including Jayashree and Malarvizhi's (2021) focus on smart microgrids and Kumar et al.'s (2022) investigation into distributed grid monitoring. These studies underline

TABLE 1.1
Result analysis

Author(s) & Year	Focus Area	Objective/Key Findings	Energy Type/ AI Technique Used	Optimization Target/Data Used	IoT or Blockchain Integration/ System Scale	Limitation or Challenge/Main Benefit	Geographic Focus/ Renewable Integration Type	Industrial or Residential/ Smart Grid Impact	Sustainability Aspect/Future Research Direction
Adewumi et al. (2024)	Energy Efficiency & Management	Improve efficiency/ AI improved energy allocation & efficiency	Mixed/ML & ANN	Resource allocation/ Operational data	No/National	Data quality limits/Higher efficiency	Africa/Hybrid systems	Industrial/Grid reliability	Energy savings/ Develop hybrid models
Ahmad et al. (2024)	Predictive Maintenance	Predict failures/ Reduced downtime via predictive models	Solar/Wind/ ML	Maintenance scheduling/ Sensor data	No/Industrial	Model interpretability/ Cost reduction	Nigeria/Solar	Industrial/ Stability improvement	Lifecycle extension/ Real-time AI models
Alam et al. (2022)	Microgrid Energy Management	Optimize scheduling/ Improved dispatch & balance	Solar + BESS/ Deep Learning	Energy scheduling/ Smart meter data	No/Residential	Large data need/ Load balancing	Korea/PV + BESS	Residential/ Local grid control	Efficiency /Transfer learning
Banik (2021)	Computational Intelligence	Power modernization/ Intelligent automation of grid	Power/Fuzzy Logic, ML	Control optimization/ Power grid data	No/Regional	Legacy compatibility/ Better automation	India/Grid	Industrial/ Automation	Rural electrification/ Cybersecurity focus
Bedi et al. (2022)	AI/IoT in Smart Cities	Smart management/ Real-time monitoring & optimization	Mixed/AI + IoT	Demand response/ IoT sensors	Yes/Urban	Interoperability issues/Urban optimization	India/Solar, Wind	Urban/Adaptive control	Smart infrastructure/ Data privacy protocols
Bouquet et al. (2024)	Solar Energy Forecasting	Forecast optimization/ Reduced error rates	Solar/Neural Networks	Forecast accuracy/ Weather data	No/Utility	Idealized assumptions/ Grid stability	Europe/Solar	Utility/Grid optimization	Forecast-based control/Weather-adaptive models

Camacho et al. (2024)	AI in Smart Cities	Support decision-making/ Improved demand prediction	Mixed/ML & Data Analytics	Energy distribution/ City-level data	Yes/Urban	Privacy & scalability/ Demand-side optimization	Mexico/Hybrid	Urban/Dynamic load control	Green city design/ Interoperable IoT AI
Chatterjee & Dethlefs (2022)	Renewable Transition	Support transition/AI accelerates RE adoption	Mixed/NLP & ML	Policy guidance/ Policy & tech data	No/Regional	Policy gaps/Faster transition	UK/Wind/Solar	National/ Integration planning	Sustainability strategy/ Explainable AI policy tools
Cicceri et al. (2023)	Cyber-Physical Energy Systems	Distributed control/ Enhanced autonomy	Renewable communities/ DL	Local optimization/ IoT data	Yes/Community	Cyber risks/Local control	Italy/Solar/Wind	Local/ Decentralized grids	Resilience /AI cybersecurity
Farrar et al. (2023)	Wind Turbine Control	Optimize control/ Efficient turbine control	Wind/ML & RL	Control parameters/ Sensor data	No/Industrial	Model generalization/ Reduced losses	USA/Wind	Industrial/Smart turbine	Efficiency / Transferable RL models
Gao et al. (2023)	Green Energy Cloud	Energy virtualization/ AI-enabled taxonomy	Mixed/AI + Cloud	Cloud services/ Cloud data	Yes/Global	Network latency/ Virtual energy mgmt.	Global/Cloud grids	Industrial/Cloud optimization	Decentralized cloud grid/Edge-AI solutions
Gnana Swathika et al. (2023)	IoT in Renewable Systems	Smart grid creation/ Real-time control	Solar/Wind/IoT & AI	Grid intelligence/ IoT data	Yes/Regional	Integration cost/ Smart monitoring	India/Mixed	Urban/Smart grid resilience	Eco-efficiency/ Scalable IoT AI
Gunavathi & Karthikeyan (2023)	Rural Electrification	Modernize infrastructure/ Improved rural access	Power/AI	System reliability/ Grid data	No/Rural	Tech adaptation/ Access expansion	India/Grid	Rural/Power delivery	Rural sustainability/ AI-assisted control
Heo et al. (2024)	Industrial Decarbonization	Techno-economic optimization/ Reduced emissions	Petrochemical/ Generative AI	Decarbonization/ Industrial data	No/Industrial	Data heterogeneity/ Emission control	Korea/ multi-energy	Industrial/Carbon neutrality	Industry 5.0/Scalable AI frameworks
Hua et al. (2022)	Blockchain for Prosumers	Enable prosumers/ Secure energy trading	Smart Grid/ Blockchain + AI	Decentralized mgmt./Energy trading data	Yes/Smart grid	Blockchain latency/Security, privacy	China/Grid	Urban/ Transparent exchange	Trust & traceability/ Lightweight blockchain
Jiao (2020)	Smart Grid Applications	Enhance automation/ Improved grid control	Grid/AI Control	Grid optimization/ Real-time data	No/Utility	Data integration issues/ Automation	China/Grid	Utility/Control accuracy	Smart energy/ Real-time AI agents

(Continued)

TABLE 1.1 (Continued)

Author(s) & Year	Focus Area	Objective/ Key Findings	Energy Type/AI Technique Used	Optimization Target/Data Used	IoT or Blockchain Integration/ System Scale	Limitation or Challenge/ Main Benefit	Geographic Focus/ Renewable Integration Type	Industrial or Residential/ Smart Grid Impact	Sustainability Aspect/Future Research Direction
Kumari et al. (2020)	Energy Cloud Management	Secure cloud systems/Efficient data mgmt.	Mixed/ Blockchain + AI	Secure operation/ Cloud energy data	Yes/Global	Computational cost/Trust enhancement	Global/Cloud/ Grid	Industrial/ Security	Green cloud/ Consensus improvement
Liu et al. (2022)	Large-Scale AI Integration	Multi-system integration/ Improved optimization	Multi-energy/ DL, ML	Carbon neutrality/ multi-source data	No/National	Data heterogeneity/ multi-system mgmt.	China/Hybrid	Industrial/ Integration	Sustainable transition/Hybrid AI-physics models
Serban & Lytras (2020)	Smart Energy Policy	Policy innovation/ Smart energy policy model	Europe/AI + Analytics	Urban efficiency/ EU data	No/Regional	Regulatory gaps/ Smart governance	Europe/Mixed	Regional/Smart city networks	Policy-driven sustainability/ Explainable policy AI
Sayed-Mouchaweh (2020)	Scalable AI Transition	Decision-support tools/AI-based planning	Energy systems/AI	Scalable optimization/ Simulation data	No/Global	Operator skills gap/Transition support	Global/Mixed	Industrial/ Automation	Green digitalization/ Human-in-loop AI

AI's contribution to enabling demand response, islanding capabilities, and efficient energy scheduling. Additionally, IoT-enabled AI frameworks for smart cities and microgrids have been gaining traction, facilitating real-time monitoring and control (Bedi et al., 2022; Swathika et al., 2023).

AI's application in predictive maintenance using anomaly detection and sensor data analytics has shown promise in improving the reliability and lifespan of renewable energy assets. Ahmad et al. (2024) and Onwusinkwue et al. (2024) provide comprehensive reviews of these developments, identifying challenges related to data quality and the urgent need for standardized evaluation protocols.

Emerging AI paradigms such as federated learning, edge AI, and digital twins are gaining importance for privacy preservation and decentralized decision-making in smart grids. Cicceri et al. (2023) and Hua et al. (2022) explore these architectures, emphasizing their scalability and resilience benefits. Explainable AI (XAI) remains an open research domain essential for fostering operator trust and meeting regulatory requirements (Liu et al., 2022; Nagpal et al., 2023).

The literature also draws attention to the socio-economic and regulatory challenges surrounding AI adoption in renewable energy. Pahle et al. (2016) discuss policy experiences influencing deployment, while Serban and Lytras (2020) and Kulkarni et al. (2023) examine the ethical, standardization, and governance dimensions of AI-driven energy systems. Issues of data privacy, cybersecurity threats, and human-in-the-loop mechanisms are recognized as critical for sustainable AI integration (Vijayalakshmi et al., 2023; Chatterjee and Dethlefs, 2022).

A notable gap in existing research lies in the limited real-world deployment of AI models beyond simulated environments. Many algorithms still lack robustness under operational constraints such as latency, hardware limitations, and imperfect data, as observed by Chatterjee and Dethlefs (2022) and Farrar et al. (2023). This gap underscores the necessity for reproducibility, open-source benchmarks, and field-level validation (Moraliyage et al., 2023; Rout and Samantaray, 2023).

1.6 CONCLUSION

The synthesis of AI techniques in renewable energy systems marks a significant advancement toward sustainable, reliable, and efficient energy infrastructures. AI's ability to enhance forecasting accuracy, optimize energy management, and enable predictive maintenance is well-documented across diverse applications. However, real-world adoption remains constrained by challenges, including data availability, model interpretability, system scalability, and compliance with regulatory standards.

Future research should prioritize developing standardized benchmarks, reproducible methodologies, and explainable AI frameworks to facilitate trust and adoption. Furthermore, integrating socio-technical considerations such as data privacy, cybersecurity, and human oversight will be critical for large-scale deployment. Continued collaboration between AI researchers, energy experts, and policymakers is essential to realize the full potential of AI in achieving global renewable energy goals.

REFERENCES

Adewumi, A., Okoli, C. E., Usman, F. O., Olu-lawal, K. A. & Soyombo, O. T. (2024). Reviewing the impact of AI on renewable energy efficiency and management. *International Journal of Science and Research Archive*, 11(1), 1518–1527. https://doi.org/10.30574/ijsra.2024.11.1.0245

Alam M.M., Hossain M.J., Habib M.A., Arafat M.Y. Hannan M.A., (2025). Aritificial intelligence integrated grid systems: Technologies, potential frameworks, challenges, and research directions. *Renewable and Sustainable Energy Reviews, 211*, 115251. https://doi.org/10.1016/j.rser.2024.115251

Alam, M. M., Rahman, M. H., Ahmed, M. F., Chowdhury, M. Z., & Jang, Y. M. (2022). Deep learning based optimal energy management for photovoltaic and battery energy storage integrated home micro-grid system. *Scientific Reports*, *12*(1), 15133. https://doi.org/10.1038/s41598-022-19147-y

Banik, A. (2021). Scope and integration of computational intelligence in traditional power sector. *Computational Intelligence and Machine Learning*, 2(1), 33–39. https://doi.org/10.36647/ciml/02.01.a005

Bedi, P., Goyal, S. B., Rajawat, A. S., Shaw, R. N., & Ghosh, A. (2022). Application of AI/IoT for smart renewable energy management in smart cities. In: Piuri, V., Shaw, R.N., Ghosh, A., & Islam, R. (eds) *AI and IoT for Smart City Applications. Studies in Computational Intelligence* (Vol. 1002). Springer, Singapore. https://doi.org/10.1007/978-981-16-7498-3_8

Bouquet, P., Jackson, I., Nick, M., & Kaboli, A. (2024). AI-based forecasting for optimised solar energy management and smart grid efficiency. *International Journal of Production Research*, 62(13), 4623–4644. https://doi.org/10.1080/00207543.2023.2269565

Camacho, J. de J., Aguirre, B., Ponce, P., Anthony, B., & Molina, A. (2024). Leveraging artificial intelligence to bolster the energy sector in smart cities: A literature review. *Energies,* 17(2), 353. https://doi.org/10.3390/en17020353

Chatterjee, J., & Dethlefs, N. (2022). Facilitating a smoother transition to renewable energy with AI. *Patterns,* 3(6), 100528. https://doi.org/10.1016/j.patter.2022.100528

Cicceri, G., Tricomi, G., D'Agati, L., Longo, F., Merlino, G., & Puliafito, A. (2023). A deep learning-driven self-conscious distributed cyber-physical system for renewable energy communities. *Sensors*, *23*(9), 4549. https://doi.org/10.3390/s23094549

Farrar, N. O., Ali, M. H., & Dasgupta, D. (2023). Artificial intelligence and machine learning in grid connected wind turbine control systems: A comprehensive review. *Energies,* 16(3), 1530. https://doi.org/10.3390/en16031530

Gao, J., Wu, J., Liu, J., & Aremanda, V. V. (2023). Green energy cloud -taxonomy, infrastructure, platform, and services. *Proceedings -17th IEEE International Conference on Service-Oriented System Engineering, SOSE 2023*. IEEE. https://doi.org/10.1109/SOSE58276.2023.00029

Gnana Swathika, O. v., Karthikeyan, K., & Padmanaban, S. (2023). *IoT and Analytics in Renewable Energy Systems (Volume 1): Sustainable Smart Grids and Renewable Energy Systems*, pp. 334. CRC Press: Boca Raton. ISBN 9781032362823. https://doi.org/10.1201/9781003331117

Gunavathi, R., & Karthikeyan, G. (2023). Modernization of rural electric infrastructure. In Vijayalakshmi, S.S., Balusamy, B., & Dhanaraj, R.K. (eds) *AI-Powered IoT in the Energy Industry. Power Systems*. Springer, Cham. https://doi.org/10.1007/978-3-031-15044-9_11

Hamdan, A., Ibekwe, K. I., Ilojianya, V. I., Sonko, S., & Etukudoh, E. A. (2024). AI in renewable energy: A review of predictive maintenance and energy optimization. *International Journal of Science and Research Archive*, 11(1), 718–729. https://doi.org/10.30574/ijsra.2024.11.1.0112

Harry, A. (2023). Unlocking synergy: Solar cell integration with radio waves for AI, IT, and business advancements. *International Journal of Multidisciplinary Sciences and Arts*, 2(2), 188–196. https://doi.org/10.47709/ijmdsa.v2i2.3211

Heo, S. K., Byun, J., Ifaei, P., Ko, J., Ha, B., Hwangbo, S., & Yoo, C. K. (2024). Towards mega-scale decarbonized industrial park (Mega-DIP): Generative AI-driven techno-economic and environmental assessment of renewable and sustainable energy utilization in petrochemical industry. *Renewable and Sustainable Energy Reviews*, 189, 113933. https://doi.org/10.1016/j.rser.2023.113933

Hua, W., Chen, Y., Qadrdan, M., Jiang, J., Sun, H., & Wu, J. (2022). Applications of blockchain and artificial intelligence technologies for enabling prosumers in smart grids: A review. In *Renewable and Sustainable Energy Reviews,* 161, 112308. https://doi.org/10.1016/j.rser.2022.112308

Jayashree, S., & Malarvizhi, K. (2021). AI-based smart micro grids. In Al-Turjman, F., Gowthaman, & N. (eds) *Advanced Controllers for Smart Cities. EAI/Springer Innovations in Communication and Computing*, pp. 1–14. Springer, Cham. https://doi.org/10.1007/978-3-030-48539-9_1

Jiao, J. (2020). Application and prospect of artificial intelligence in smart grid. *IOP Conference Series: Earth and Environmental Science*, 510(2), 022012. https://doi.org/10.1088/1755-1315/510/2/022012

Kulkarni, G. R., Tamta, M. K., Kumar, A., Nomani, M. Z. M., Singh, C., & Pallathadka, H. (2023). The role of artificial intelligence (AI) in creating smart energy infrastructure for the next generation and protection climate change. *Smart Innovation, Systems and Technologies, 290.* Springer, Singapore. https://doi.org/10.1007/978-981-19-0108-9_47

Kumar, R. S., Saravanan, S., Pandiyan, P., & Tiwari, R. (2022). Impact of artificial intelligence techniques in distributed smart grid monitoring system. In Padmanaban, S., Holm-Nielsen, J.B., Padmanandam, K., Dhanaraj, R.K., & Balusamy, B., (eds) *Smart Energy and Electric Power Systems: Current Trends and New Intelligent Perspectives.* Elsevier. https://doi.org/10.1016/B978-0-323-91664-6.00005-X

Kumari, A., Gupta, R., Tanwar, S., & Kumar, N. (2020). Blockchain and AI amalgamation for energy cloud management: Challenges, solutions, and future directions. *Journal of Parallel and Distributed Computing*, 143, 148–166. https://doi.org/10.1016/j.jpdc.2020.05.004

Liu, Z., Sun, Y., Xing, C., Liu, J., He, Y., Zhou, Y., & Zhang, G. (2022). Artificial intelligence powered large-scale renewable integrations in multi-energy systems for carbon neutrality transition: Challenges and future perspectives. *Energy and AI*, 10, 100195. https://doi.org/10.1016/j.egyai.2022.100195

Mohana, M., Saidi, A. S., Alelyani, S., Alshayeb, M. J., Basha, S., & Anqi, A. E. (2021). Small-scale solar photovoltaic power prediction for residential load in Saudi Arabia using machine learning. *Energies, 14*(20), 6759. https://doi.org/10.3390/en14206759

Moraliyage, H., Haputhanthri, D., Samarajeewa, C., Mills, N., de Silva, D., Manic, M., & Jennings, A. (2023). Automated machine learning in critical energy infrastructure for net zero carbon emissions. *IEEE International Symposium on Industrial Electronics*, 2023-*June*. IEEE. https://doi.org/10.1109/ISIE51358.2023.10227985

Nagpal, N., Alhelou, H. H., Siano, P., Padmanaban, S., & Lakshmi, D. (2023). *Applications of Big Data and Artificial Intelligence in Smart Energy Systems* (Vol. 2). CRC Press/Taylor Francis, 250 pages. ISBN 9788770228275. https://doi.org/10.1201/9781003440864

Onwusinkwue, S., Osasona, F., Ahmad, I. A. I., Anyanwu, A. C., Dawodu, S. O., Obi, O. C., & Hamdan, A. (2024). Artificial intelligence (AI) in renewable energy: A review of predictive maintenance and energy optimization. *World Journal of Advanced Research and Reviews*, 21(1), 2487–2499. https://doi.org/10.30574/wjarr.2024.21.1.0347

Pahle, M., Pachauri, S., & Steinbacher, K. (2016). Can the Green Economy deliver it all? Experiences of renewable energy policies with socio-economic objectives. *Applied Energy, 179*, 1331–1341. https://doi.org/10.1016/j.apenergy.2016.06.073

Rout, S., & Samantaray, S. (2023). Smart technologies for a sustainable future: IoT and AI in renewable energy. *2023 IEEE 3rd International Conference on Applied Electromagnetics, Signal Processing, and Communication, AESPC 2023*. IEEE. https://doi.org/10.1109/AESPC59761.2023.10390251

SaberiKamarposhti, M., Kamyab, H., Krishnan, S., Yusuf, M., Rezania, S., Chelliapan, S., & Khorami, M. (2024). A comprehensive review of AI-enhanced smart grid integration for hydrogen energy: Advances, challenges, and future prospects. *International Journal of Hydrogen Energy*, 67, 1009–1025. https://doi.org/10.1016/j.ijhydene.2024.01.129

Sahoo, S. (2022). Power and energy management in smart power systems. In Padmanaban, S., Holm-Nielsen, J.B., Padmanandam, K., Dhanaraj, R.K., & Balusamy, B., (eds) *Artificial Intelligence-based Smart Power Systems*, pp. 349–375. Wiley – IEEE Press. https://doi.org/10.1002/9781119893998.ch18

Sayed-Mouchaweh, M. (2020). *Artificial Intelligence Techniques for a Scalable Energy Transition: Advanced Methods, Digital Technologies, Decision Support Tools, and Applications*. Springer, Cham. https://doi.org/10.1007/978-3-030-42726-9

Serban, A. C., & Lytras, M. D. (2020). Artificial intelligence for smart renewable energy sector in Europe - Smart energy infrastructures for next generation smart cities. *IEEE Access, 8,* 77364–77377. https://doi.org/10.1109/ACCESS.2020.2990123

Sharma, S. P., & Yadav, D. K. (2023). Renewable energy systems energy modeling using deep learning techniques. *2023 2nd International Conference for Innovation in Technology, INOCON 2023*. IEEE. https://doi.org/10.1109/INOCON57975.2023.10101286

Stecuła, K., Wolniak, R., & Grebski, W. W. (2023). AI-driven urban energy solutions from individuals to society: A review. *Energies,* 16(24), 7988. https://doi.org/10.3390/en16247988

Swathika, O. V. G., Karthikeyan, K., & Padmanaban, S. (2023). *IoT and Analytics in Renewable Energy Systems (Volume 2): AI, ML and IoT Deployment in Sustainable Smart Cities*, pp. 344. CRC Press, Boca Raton. ISBN 9781032448299. https://doi.org/10.1201/9781003374121

Tran, T. A. (2017). A research on the energy efficiency operational indicator EEOI calculation tool on M/V NSU JUSTICE of VINIC transportation company, Vietnam. *Journal of Ocean Engineering and Science,* 2(1), 55–60, (SCIE, IF = 13.0). https://www.sciencedirect.com/science/article/pii/S2468013316300614.

Tran, T. A. (2021). Comparative analysis on the fuel consumption prediction model for bulk carriers from ship launching to current states based on sea trial data and machine learning technique. *Journal of Ocean Engineering and Science,* 6(4), 317–339 (SCIE, IF = 13.0), https://www.sciencedirect.com/science/article/pii/S2468013321000206

Vijayalakshmi, S., Savita, & Durgadevi, P. (2023). AI and IoT in improving resilience of smart energy infrastructure. In Vijayalakshmi, S.S., Balusamy, B., & Dhanaraj, R.K. (eds) *AI-Powered IoT in the Energy Industry. Power Systems*. Springer, Cham. *Power Systems*. https://doi.org/10.1007/978-3-031-15044-9_9

2 AI-assisted Wave Energy Converters

Technology Evolution, Applications, and Future Perspectives

Ali Azam, Ammar Ahmed, Attia Qammar, Ayesha Asghar, and Tien Anh Tran

2.1 OVERVIEW

This chapter provides a concise yet comprehensive discussion of various AI-assisted applications in modern wave energy converters (WECs). Recent research trends, hotspots, innovation horizons, and frontiers were determined through scientometric simulations. Also, the self-powered and self-sensing applications of modern WEC systems are identified. Finally, the emerging challenges and future perspectives are discussed to assist a robust deployment of next-generation AI-driven WEC systems. The principal objectives of this chapter are:

i. To highlight the applications and advancements of AI across various departments of WECs.
ii. To consolidate emerging knowledge structures in AI-driven WEC research, using bibliometric evidence to demonstrate technological pathways.
iii. To identify emerging trends, innovation horizons, and critical challenges in AI-driven WECs, thereby informing future research, commercialization strategies, and policy frameworks.

2.2 BACKGROUND AND CURRENT STATUS

Overall, the WECs have been generally classified into oscillating water columns (OWCs), oscillating body (OB-WECs), surge converters, and overtopping devices (OT-WECs) [1, 2], as shown in Figure 2.1.

The OWCs, such as the backwater bent buoy [3] and the Spar buoy [4], use the rise and fall of water within a chamber to compress and decompress air, which drives a turbine to generate electricity [5]. The OT-WECs, such as the wave dragon [6], capture wave water that overtops into a reservoir, and the stored water is then released

DOI: 10.1201/9781003671039-2

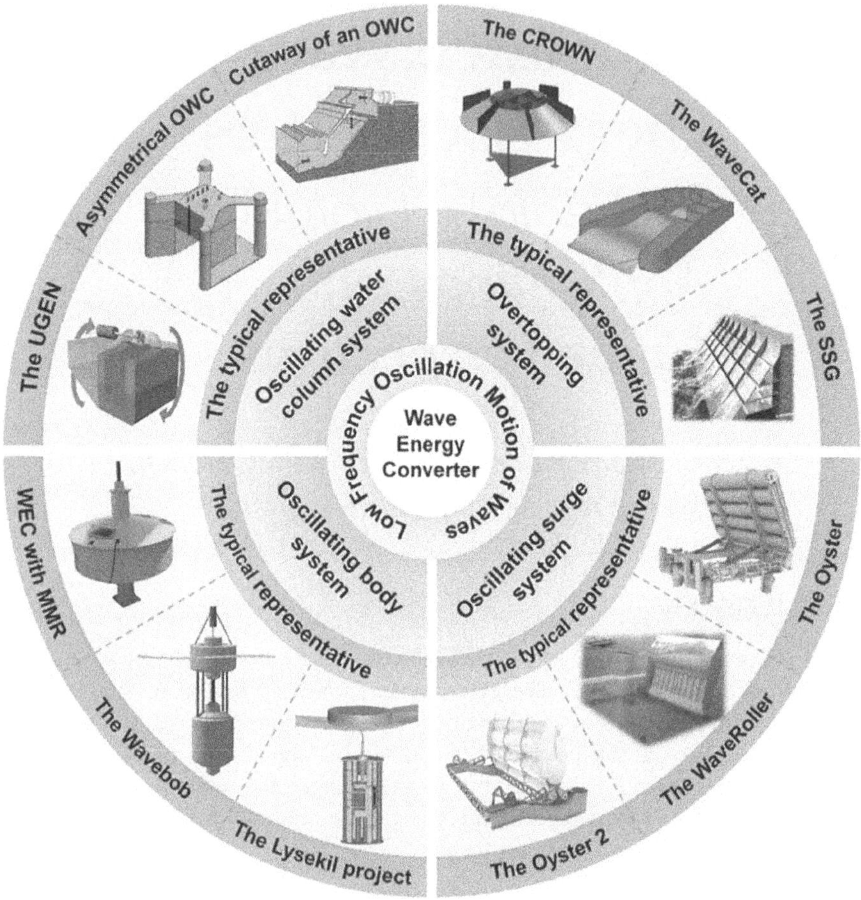

FIGURE 2.1 WEC categorization demonstrated by Li et al. [2] (adapted with the permission of the publisher).

to flow back to the ocean through turbines, generating electricity. The OB-WECs usually consist of two parts: (i) a floater or oscillator that captures wave motions along various directions and (ii) a reactive part or power take-off (PTO) mechanism to convert the kinetic energy of the oscillator into electrical power.

Figure 2.2 shows the recent research trends of different WEC categories in recent years (based on [7]). It shows that the OB-WECs are the most studied systems because they can harvest energy from various wave motions through various designs with a diverse range of operational characteristics [8], for instance, the Oyster, Pelamis, and Searev [9]. The OB-WECs can be further categorized as: (i) the point absorbers (PA-WECs) exploiting the heave motion, (ii) terminators harvesting energy from the surge component, and (iii) attenuators exploiting the pitch motion to extract wave energy. Approximately 53% of the developed WECs are point absorbers, 33% are categorized as terminators, and the remaining 14% are attenuators [10]. Due to their compact design, versatility, simple structure, and power performance [11], the

Author affiliation	Overtopping devices					Oscillating water columns					Oscillating bodies				
	2019	2020	2021	2022	2023	2019	2020	2021	2022	2023	2019	2020	2021	2022	2023
China	9	16	18	17	33	256	353	471	620	754	504	680	856	1038	1382
Indonesia	-	-	-	-	-	-	3	2	2	4	5	4	4	12	8
South Korea	6	-	2	1	2	28	27	33	32	47	74	77	88	71	91
Japan	2	2	1	2	2	30	40	44	61	57	78	100	115	119	118
Malaysia	3	-	-	-	1	11	5	11	13	10	19	16	24	23	31
Singapore	1	1	1	2	-	9	24	18	23	17	36	52	38	43	39
Thailand	-	-	-	-	-	2	3	6	8	3	3	6	6	9	15
Vietnam	1	-	-	1	-	2	3	4	4	2	4	10	9	18	6
Canada	-	5	1	-	3	48	65	65	78	77	96	96	115	121	180
Germany	-	-	-	-	1	110	88	108	112	126	223	222	232	245	267
France	3	1	1	-	1	95	105	99	110	121	165	158	169	218	218
Italy	3	2	8	8	3	81	95	92	82	102	163	177	200	173	184
Netherlands	1	1	-	3	2	27	32	28	41	31	62	70	60	79	54
Spain	9	9	14	8	1	121	96	116	114	127	131	131	169	159	170
UK	5	9	14	4	11	78	90	99	143	121	170	214	196	219	246
USA	5	5	7	4	3	211	230	219	278	280	410	393	443	458	518

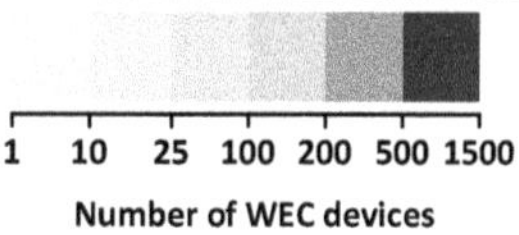

FIGURE 2.2 Statistics on WEC devices studied recently (based on statistics found in [7]).

PA-WECs have played a significant role in self-sensing and self-powered sustainable sea applications. However, their advancement is hindered by challenges related to maintenance, high energy costs [12], reliability [13], ecological risk assessment, survivability, peak shaving [14], and ocean biofouling [15, 16]. Based on their sizes and output power capacities, WECs can be classified into macro- and micro-scale WECs.

2.3 AI-ASSISTED ADVANCEMENTS IN WECS

The integration of AI and Internet of Things (IoT) into WEC systems represents a paradigm shift in the way wave energy is harnessed. By leveraging AI-driven algorithms and IoT-enabled sensors, it is possible to enhance the accuracy of wave resource assessments [17, 18], optimize the geometry and design of floaters [19], and improve the efficiency of PTO mechanisms [20]. Furthermore, AI and IoT can play a critical role in assessing and mitigating the ecological risks associated with WEC deployments [21, 22], ensuring that these technologies align with broader sustainability objectives.

Many real-time WEC applications have considered using deep learning and AI to predict wave climate data, WEC performance, and control strategies to exploit wave resources better. Gu et al. [23] summarized that AI could influence wave energy identification, exploration, forecasting, harvesting, capturing, and management. The AI-based algorithms could be categorized into different echelons, for example:

i. **Device** – fault diagnosis, control, monitoring, and prediction of the device's life cycle,
ii. **Operation** – forecasting waves and prediction of instability and power generation,

iii. **Energy plant level** – resource allocation, ride-through and limited life distribution, site selection, forecasting output, and
iv. **Power dispatching** – reactive power management and distribution, voltage and frequency regulation.

Overall, the AI applications in research, development, and deployment (RD&D) of the recent WECs can be classified into the following sub-domains, as shown in Table 2.1.

TABLE 2.1
Applications of AI in various sub-domains of recent WECs

Sr. No.	Application domains of AI in WECs	Breakthrough advancements	References
1	Wave resource assessment	AI, ocean digital twins, deep brief networks, long short-term memory (LSTM), recurrent neural networks, deep hybrid models, convolution neural networks (CNN), gated recurrent networks (GRU), and deep neural networks (DNN) used in oceanography as deterministic wave forecasting models	[24–27],
2	Buoy geometry and WEC design optimization	Back propagation ANN model to optimize the geometry of a point absorber buoy with random Bessel control points to find the modified optimal versions of hemispherical, conical, and cylindrical buoys, machine learning algorithm for design optimization of the WEC.	[1, 5, 28–30, 31, 32]
3	Ecological risk assessment of WEC projects	Development of a web-based tool (WEC-ERA).	[33–36]
4	WEC arrays, wave farms, and offshore parks	Machine learning (ML)-based array layout optimization algorithms, genetic algorithms to attain the highest interaction factors, ANN algorithm for optimizing the WEC array layouts.	[37–40]
5	Assisting IoT through portable PTO	Model-free ANN algorithm to facilitate finding the optimal PTO damping and stiffness, machine learning in maximizing the wave power extraction of a WEC.	[41–43]

(Continued)

TABLE 2.1
(continued)

Sr. No.	Application domains of AI in WECs	Breakthrough advancements	References
6	Scalability of the WECs	A self-stabilizing scalable PA-WEC, an S-shaped scalable WEC.	[1, 44]
7	Life cycle thinking (LCA) of WECs	Dynamic nature of AI in energy sustainability, the complex maintenance of baseline assessments.	[45–47]
8	AI-enabled WEC control	Non-casual WEC control algorithm using a feedforward ANN, model predictive control, constant and time-varying damping control, reactive control, and latching/unlatching control.	[48–50]
9	AI-powered wave-to-wire models	A W2W model based on deep reinforcement learning (DRL), an AI-assisted W2W model.	[51–53]
10	Wave-structure interaction, viscous drag, moorings, and non-linear hydrodynamics	The ML-based models and neural network trained to predict drag coefficients, a Bi-LSTM neural network to forecast mooring tension.	[54, 55]

2.4 TRACING THE DEVELOPMENTAL ARC OF AI-ASSISTED PA-WECS

In this section, the recent advances in AI-assisted WECs are mapped through a bibliometric study using a Keywords Cluster-driven Knowledge Structuring (KCKS) model to determine their technological progression pathways, challenges, and contemporary knowledge landscape.

2.4.1 The Methodological Approach

Figure 2.3 represents the strategy used to map the recent advances in AI-assisted WECs through the bibliometric networks.

The study is based on the following methods: (i) software definition–CiteSpace and VOS viewer were selected for bibliographic simulations followed by data processing, (ii) the databases within Web of Science Core Collection (WOSCC), offering compatible file formats, were identified, (iii) using an optimal keyword code to converge the wider WEC domain into a narrower AI-assisted WEC domain, (iv) filtration, and retrieval of the data with full bibliographic details, (v) bibliometric

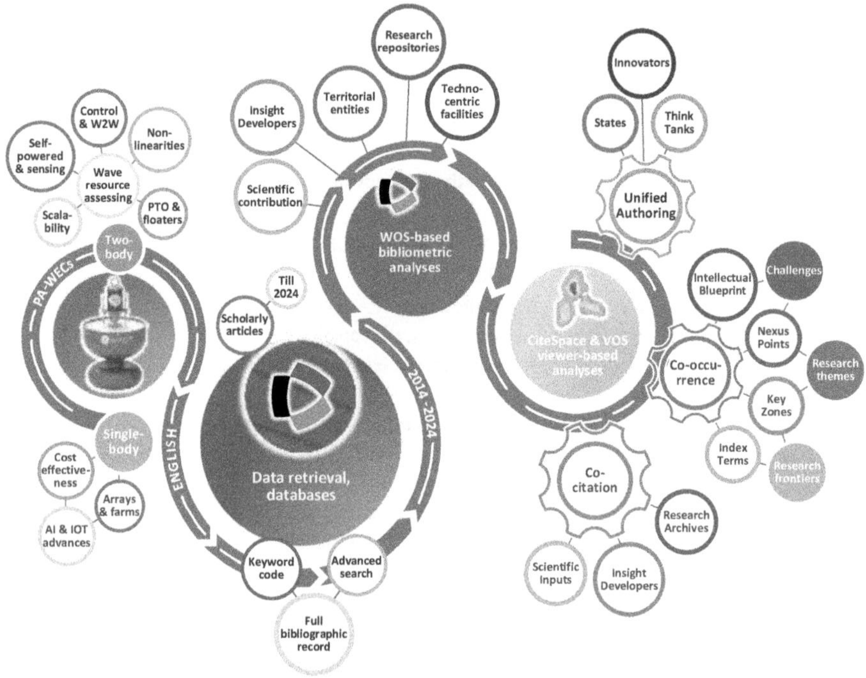

FIGURE 2.3 The schematic of tracing the developmental arc of AI-assisted WECs.

simulations, (vii) refining, filtering, and derivation of authentic statistics, and (viii) post-processing using the KCKS model.

2.4.1.1 Pre-processing for Bibliometric Analysis

Full-length articles published between 2000 and 2024 on the AI-assisted WECs were obtained from the WOSCC databases, including Science Citation Index Expanded and Emerging Sources Citation Index. The iteratively-developed code, "((((ALL=("wave energy converter" OR "wave energy harvester" OR "wave energy extraction" OR "wave power generation" OR "WEC array" OR "wave farms" OR "WEC design" OR "Wave parks" OR "PA-WEC array" OR "PA-WEC control" OR "Buoy design" OR "wave resource assessment")) AND ALL=("point absorber" OR "heave" OR "heaving" OR "PA-WEC" OR "heaving WEC" OR "smart sea applications" OR "wave to wire" OR "control" OR "power enhancement" OR "power performance" OR "maximization" OR "optimization" OR "design" OR "power take-off" OR "PTO" OR "wave resource assess*" OR "array layout optimiz*")) AND ALL=("artificial intelligence" OR "machine learning" OR "deep learning" OR "data driven" OR "ANN" OR "AI" OR "LSTM" OR "neural network" OR "WEC control" OR "wave to wire" OR "IoT" OR "Internet of things"))" was used to collect a total of 166 most relevant publications. The patents, proceedings papers, editorial materials, book reviews, letters, and biographical items were eliminated, and only seven review and 159 original research articles were exported as plain text and tab-delimited files with full bibliographic details.

2.4.1.2 Bibliometric Analysis in CiteSpace and VOS Viewer

The exported publications with full bibliographic details were imported as plain text files in CiteSpace (6.3.R1) and the tab-delimited files in VOS viewer (1.6.18) software, and the following settings were designated for bibliometric simulations, as shown in Table 2.2.

While conducting the simulations the following entities were considered including scientific contributions, insight developers, territorial entities, research repositories or archives, techno-centric facilities, innovators, nexus points, and indexing terms to identify the challenges, research themes, and frontiers in the AI-assisted WECs. Two parameters, (i) burst resilience and (ii) network mediation index or centrality, were used to highlight the items in the network. The subject is represented by an "item" in the pattern. Items are connected through links of particular width and length. The betweenness centrality is the relation between the shortest link between two subjects and the total of all the shortest links, as given below [56–58],

$$\text{Centrality}(\text{node}_n) = \sum\nolimits_{i \neq j \neq k} \frac{M_{jk}(i)}{M_{jk}} \qquad \text{Eqn. (1)}$$

here M_{jk} is the cumulative sum of the shortest links between the items j and k; $M_{jk}(i)$ is the total number of links connecting the subject i.

2.4.1.3 KCKS Model

Ahmed et al. [59] demonstrated that keyword-based clusters outperform document-based clusters in terms of label recency, accuracy, betweenness centrality, and overall link strength. In the keywords-based approach, the clusters are developed using highly co-occurring keywords, and in the documents-based approach, the clusters are made using highly cited documents. Both approaches were used to develop clusters in the CiteSpace and VOS viewer to validate the hypothesis, and the data were scrutinized and characterized to develop the knowledge structure using the

TABLE 2.2
Settings for bibliometric simulations

Sr.	Settings	Selections
1	Analysis type	Keyword nodes with term sources: title, abstract, and author keywords.
2	Counting method	Fractional counting.
3	Item type	Authors, Keywords, journals, institutions, and cited references.
4	Criteria of selection	Top 10% of the most cited or occurred entities.
5	Normalization	Association strength.
6	Nodes/ Links	Diamond-shaped nodes, curved links, and colored visualizations.
7	Time slicing	Between 2000 and 2024.

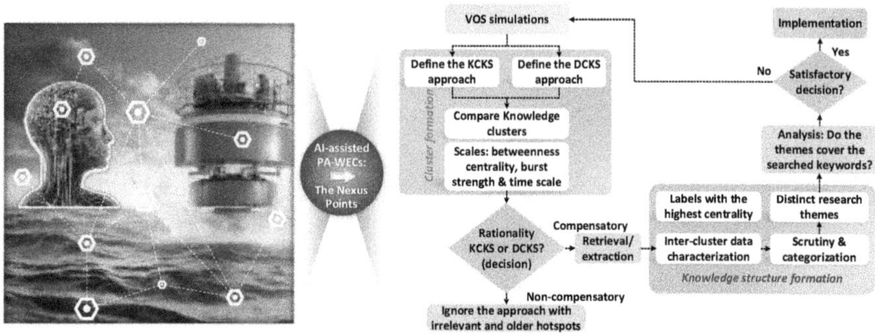

FIGURE 2.4 The framework of keyword cluster-driven knowledge structuring (KCKS) algorithm for evaluating technological evolution pathways of the AI-assisted WEC research.

algorithm shown in Figure 2.4. These methods referred to the KCKS and document cluster-driven knowledge structuring (DCKS). Later, the leading labels with the highest centrality were compared, and the results proposed in [59] were analyzed to find the accurate hypothesis. In addition to [59], this study extends the concepts of KCKS and DCKS algorithms to determine the knowledge structure of the AI-assisted WEC research. The resulting knowledge structure frameworks were compared against each other similar to the previous stage (clusters). In addition, the research themes found from KCKS and DCKS methods were compared to the searched (input) topics, and it was found that the themes obtained through the KCKS approach covered the whole range of input topics and were more authentic than those from the DCKS approach in terms of recency, betweenness centrality, and overall link strength. These results verified the preliminary outcomes revealed in [59].

2.4.2 The Developmental Arc of the AI-assisted WEC Research

2.4.2.1 Nexus Points and Key Zones

Two keywords are connected in a co-occurrence map if they appear simultaneously in a research probe or dataset. Recently published keywords were sourced to identify the nexus points or key zones (also known as hotspots) within AI-assisted WEC research. The published research on various AI-driven applications in WECs was mapped to identify precise nexus points and key zones, as shown in Figure 2.5. The keywords interconnected in Figure 2.5 demonstrate strong relevance to the selected domain of "AI-driven WECs"," as represented by the search code used for data retrieval. This validates the effectiveness of the search code, and the convergence method employed in the current study. Table 2.3 enlists top 30 keywords with the strongest citation bursts related to AI-assisted WECs from 2014 to 2024.

2.4.2.2 Think Tanks, Archives, and Scientific Insights

Co-citation is a link between two insight developers, contributions, and repositories or archives, with simultaneous academic attributions in the same research probe. The VOS maps, constructed through co-citation analysis, were employed to pinpoint

TABLE 2.3
Top 30 keywords with the strongest citation bursts related to the AI-assisted WECs between 2014 and 2024

No.	Keywords	Year	Strength	Begin	End	2014–2024
1	Wave-to-wire model	2020	0.73	2020	2020	
2	Deep reinforcement learning	2020	0.58	2020	2020	
3	Cost of energy	2020	0.58	2020	2020	
4	Shape optimization	2021	1.2	2021	2021	
5	Economic model predictive control	2021	0.64	2021	2024	
6	Energy or power management	2021	0.6	2021	2021	
7	Frequency control	2021	0.6	2021	2021	
8	Economic optimization	2021	0.6	2021	2021	
9	Damping and causal control	2021	0.6	2021	2021	
10	Triboelectric nanogenerator	2022	1.16	2022	2022	
11	Computational intelligence techniques	2022	0.58	2022	2022	
12	Energy parks and farms	2022	0.58	2022	2022	
13	Adaptive genetic algorithm	2022	0.58	2022	2022	
14	Array layout optimization	2022	0.58	2022	2022	
15	Co-located wave	2022	0.58	2022	2022	
16	Energy or power maximization	2022	0.58	2022	2022	
17	Complex-conjugate control	2022	0.58	2022	2022	
18	Hybrid wind-WEC systems	2023	1.07	2023	2024	
19	Hydraulic PTO	2016	0.91	2023	2024	
20	Combined wave height forecasting	2023	0.53	2023	2024	
21	Buoy draft adjustment	2023	0.53	2023	2024	
22	Artificial neural networks	2023	0.53	2023	2024	

(Continued)

TABLE 2.3
(Continued)

No.	Keywords	Year	Strength	Begin	End	2014–2024
23	Electrical power production	2023	0.53	2023	2024	
24	Electric load control	2023	0.53	2023	2024	
25	Adaptive Particle Swarm Optimization	2023	0.53	2023	2024	
26	Dry test rig	2023	0.53	2023	2024	
27	Control co-design (CCD)	2023	0.53	2023	2024	
28	Capture width ratio	2023	0.53	2023	2024	
29	Energy saving	2023	0.53	2023	2024	
30	ANN MLP regression model	2023	0.53	2023	2024	

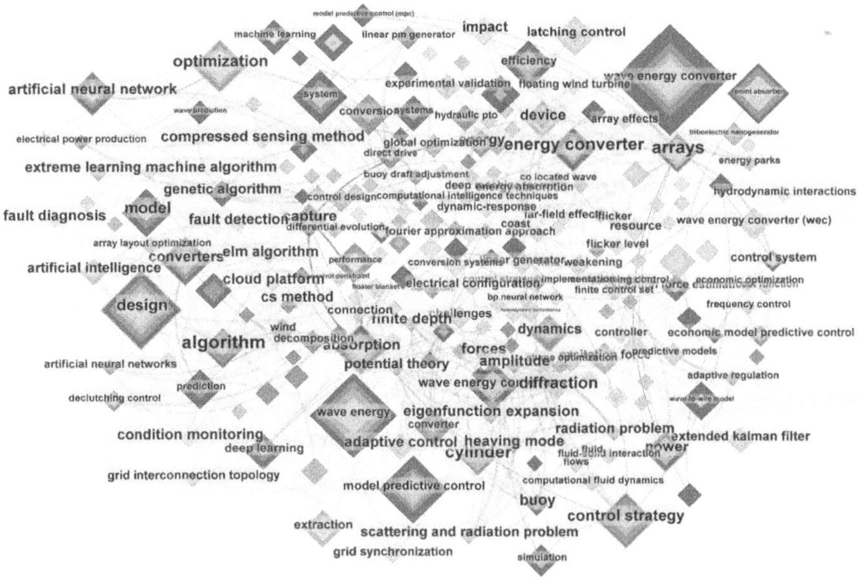

FIGURE 2.5 The most co-occurring keywords and hotspots related to the AI-assisted WECs (visualized in CiteSpace).

the leading developers, their contributions, and associated repositories in the field of AI-assisted WECs, as outlined in Table 2.4, including their aggregate link strength. The entities were organized and prioritized according to their citation metrics within the WOS databases.

2.4.2.3 Knowledge Clusters and Structure

The knowledge clusters within AI-assisted WEC research were assessed, as depicted in Figure 2.6, to analyze its multi-level knowledge topology or knowledge structure. The published data were segmented into ten unique clusters, each encompassing various emerging sub-fields that are currently under global scholarly focus. The silhouette coefficient (S), which gauges the degree of cohesion within clusters, is defined by [59],

$$S(n) = \frac{p(n) - q(n)}{\max\{p(n), q(n)\}} \qquad \text{Eqn. (2)}$$

where n represents the total number of subjects, $q(n)$ denotes the centralized disparity of subject n with other subjects within the same cluster, and $p(n)$ indicates the minimum centralized disparity of subject n with subjects in adjacent clusters. The S varies between 1 and −1; $S = 1$ indicates robust clustering, and $S = 0.5$ reveals relatively weak data clustering. In this investigation, nearly all clusters exhibited an S greater than 0.9, indicating robust data clustering.

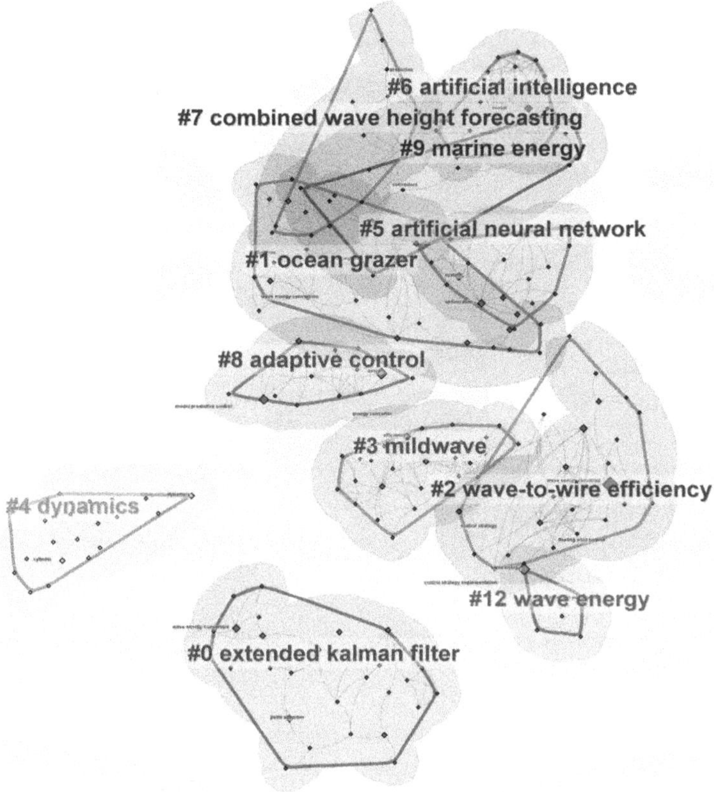

FIGURE 2.6 Insights clusters of the AI-assisted WEC research.

The leading keywords in the clusters are Ocean grazer, wave-to-wire efficiency, mild waves, non-linear hydrodynamics, extreme response and fatigue, extended Kalman filter, triboelectric nanogenerators (TENGs), artificial neural network, offshore hybrid wind-wave systems, AI, combined wave height forecasting, adaptive control, adjustable natural frequency, frequency bandwidth expansion, ecological risk assessment, wave parks or wave farms, and array layout optimization.

Knowledge structure is a multi-level knowledge topology or hierarchical map including all spheres, micro-fields, niche disciplines, research deficits, specialized topics, and uncharted zones of the relevant field or technology [57, 59] for the AI-assisted WECs. Henceforth, ten distinct thematic disciplines can be proposed, including (#0) WEC principles – state-of-the-art, classification, design, performance assessment, and maintenance, (#1) wave resource management – prediction and forecasting, (#2) non-linear hydrodynamics – moorings, viscous drag, wave-structure interactions, testing-vs-simulations (#3) offshore structures – wave-wind hybrid systems, floating platforms (#4) WEC control design and strategies – W2W models, extremities and protection (#5) multi-absorbers – arrays, farms, parks, (#6) ecological risk assessment – marine life and ecosystem preservation, (#7) self-powered and self-sensing nanogenerators – triboelectricity or piezoelectricity, (#8) ultramodern

TABLE 2.4
Leading think tanks, archives, and scientific insights (based on citations in WOS) in the AI-assisted WEC research

Sr.	Think Tanks	Attributions	Holistic Link Strength	Archives	Attributions	Holistic Link Strength	Scientific Insights	Attributions	Holistic Link Strength
1	Falnes, J	464	444.62	Renew Energ.	1545	1345.64	[60]	163	163
2	Babarit, A	331	320.38	Ocean Eng.	1487	1294.63	[61]	144	144
3	Falcao, Afd	174	173.65	Appl Ocean Res.	760	712.06	[62]	92	91
4	Penalba, M	115	111.05	Renew Sust Energ Rev.	629	594.19	[63]	75	75
5	Budal, K	109	106.22	Energies	516	483.67	[64]	71	71
6	Hals, J	107	105.32	Energy	338	322.26	[65]	69	69
7	Drew, B	105	105	Appl Energ.	298	279.03	[66]	56	56
8	Evans, Dv	105	102.49	IEEE T Sustain Energ.	281	265.01	[67]	54	54
9	Cummins, W	94	94	J Offshore Mech Arct.	253	244.16	[68]	49	49
10	Folley, M	89	85.91	Int J Mar Energy	208	204.82	[69]	49	49
11	Fusco, F	86	83.62	J Fluid Mech.	196	188.54	[70]	46	46
12	Xu, S	84	70.47	Ocean Waves Oscillat.	187	186.84	[71]	44	44
13	Salter, Sh	82	80.77	P I Mech Eng A-J Pow.	150	148	[72]	43	43
14	Newman, Jn	80	78.32	J Mar Sci Eng.	142	137.21	[73]	40	40
15	Zhang, Xt	80	75	Energ Conver. Manag.	140	133	[74]	39	39

(Continued)

TABLE 2.4
(Continued)

Sr.	Think Tanks	Attributions	Holistic Link Strength	Archives	Attributions	Holistic Link Strength	Scientific Insights	Attributions	Holistic Link Strength
16	Falcao, Afdo	72	71.02	Mar Struct.	136	134.69	[75]	39	39
17	Budar, K	71	71	Coast Eng.	134	128.12	[76]	38	38
18	Korde, Ua	70	68.96	Nature	129	127.82	[77]	38	38
19	Clement, A	67	66.76	IET Renew Power Gen.	126	123.04	[78]	37	37
20	Giorgi, G	63	59.34	IEEE T Energy Conver.	121	114.47	[79]	37	37

energy capture enhancement – multi-frequency resonance, bandwidth expansion, draft adjustment, and (#9) artificial neural networks & 5G-IoT – smart sea applications. Recent advancements in AI-assisted WEC research can be stratified into these disciplines, offering a roadmap for future researchers.

2.4.2.4 Innovation Horizons and Research Frontiers

The nexus points and key zones represent the most intensely researched keywords with peak occurrences over time, while innovation horizons or research frontiers denote the latest emerging zones or sub-fields of a research direction, currently under exploration or poised for imminent study. Through comprehensive VOS analysis of the key focus zones, 150 cutting-edge research frontiers within the AI-assisted WEC literature (published up to 2024 and ranked by network strength) have been identified and detailed in Table 2.5, contributing valuable insights for future researchers.

2.4.2.5 Operational Robustness of the KCKS Model

The following verifications were performed to examine the operational robustness of the KCKS in assessing AI-assisted PA-WECs, as shown in Table 2.6, and a comparison was made with the same study repeated with the DCKS-based approach. The knowledge structure obtained through the KCKS approach covered the whole range of input topics and frontiers and was superior to those from the DCKS approach in terms of recency, betweenness centrality, coverage of breakthrough insights, and overall link strength. It justifies the application of the KCKS algorithm in the bibliometric study of a broader range of relevant literature.

2.5 SMART SEA APPLICATIONS OF MODERN WECS

2.5.1 Micro-scale WECs and Their Applications

These are small-scale point absorber devices utilizing electromagnetic, piezoelectric, and TENGs to harvest wave energy to yield an output ranging between a few mW and Watts [80]. They are distinct in terms of design flexibility, structural versatility, and their standalone capabilities in self-sensing applications. Recent advances in micro-WECs are shown in Figure 2.7. A ball-shaped hybrid electromagnetic and piezoelectric WEC was developed by Qi et al. [81] for self-sensing and self-powered structural health monitoring of a sea-crossing bridge, as shown in Figure 2.7(a). The movement of the water wave causes the magnetic core to slide, intersecting magnetic flux lines to produce an electric current, while simultaneously causing the piezoelectric sheet to twist and generate electricity. Similarly, Gao et al. [82] demonstrated a WEC resembling a gyroscope with a TENG with an optimal wave energy capture performance and potential for self-powered smart sea gadgets, as shown in Figure 2.7(b). The output of the TENGs installed on the outside and inside of the WEC reached 160V and 730V, respectively, which was enough to run thermometer and several LED lights, justifying their feasibility for real-time self-powering applications with a demonstration of highly durable structural composition.

Figure 2.7(c) represents a bioinspired micro-WEC device resembling butterfly wings with a TENG to absorb energy from ultra-low frequency waves [83]. The

TABLE 2.5
150 cutting-edge research frontiers in the AI-assisted WECs, arranged in the descending order of their network strength in the VOS map in the following format: "Frontier (with its links in the map)"

WEC-SIM (66)	Memory effect (14)	Wavestar diameter (11)	SDGs (9)	Direct-drive PTO (7)
Wind-WEC hybrids (46)	Power set-point tracking (14)	Intermittency (11)	Coupled dynamic analysis (9)	MATLAB/Simulink (7)
Extreme mooring tension (37)	Hybrid-frequency-time domain method (14)	Smart floating cities (11)	Hybrid wave-current energy converter (9)	Model predictive control with convex (7)
Levelized cost of energy (29)	Theoretical and practical performance (14)	Sequential quadratic programming (11)	Brace less floating platform (9)	Wave flume experiments (7)
Maximizing control through ANN (26)	Prony approximation (14)	Multi-mode WEC (11)	Nylon mooring (8)	Deep reinforcement learning (7)
Artificial Neural Network (25)	Economic optimization (14)	Smart sea applications (11)	Non-linear axial stiffness (8)	Neural networks (7)
Reactive power (21)	Uppsala WEC (13)	PTO damping (11)	PA-WEC array (8)	Parametric study (6)
Resonance effect (21)	Extreme events (13)	Multi-verse optimizer (11)	Variable stiffness (8)	Interval type-2 fuzzy system (6)
Techno-economic analysis (20)	Power quality (13)	Hydrodynamic non-linearities (11)	Ropes and moorings (8)	Black hole optimization algorithm (6)
SPAR buoy (20)	Multi-objective optimization (13)	Resonant WEC (11)	Multi-body motion coupling (8)	Maximizing power capture (6)
Non-linear control (18)	Offshore Wind-WEC energy (13)	Short-term extreme response (11)	Viscous damping coefficient (8)	Bragg scattering (5)
Maximum covariance analysis (17)	Wave-to-grid control (13)	Probabilistic failure (11)	Extreme loads (8)	Adaptive kernel density estimation (5)
Hydrogen production (17)	Embedded focused waves (13)	Platform motions (11)	Array layout optimization (8)	Long-term extremes (5)
North-Atlantic oscillation (17)	Iyapunov control (13)	Long-term extreme response (11)	Artificial Intelligence in PA-WECs (8)	Pitching WEC (5)

TABLE 2.5
(Continued)

Far offshore (17)	Moordyn Plus (13)	Multi-point absorbers (11)	Dynamic axial stiffness (8)	Wave power conversion efficiency (4)
Wave focusing (17)	Energy parks (13)	Improved bistable mechanism (10)	Digital twin (8)	Operation and maintenance (4)
Prediction models (16)	Dimensional optimization (12)	Semi-enclosed moonpool (10)	OpenfoamV1906(8)	WEC bandwidth expansion (4)
Low-energy-density sea (15)	Control-oriented modeling (12)	Mechanical networks (10)	Model test (8)	Areo-servo-hydro-elastic-simulation (4)
Decarbonization (15)	Power fluctuation mitigation (12)	Surge converters (10)	Non-linear autoregressive (8)	Environmental impact (4)
Ports and harbor break waters (15)	Response surface methodology (12)	Parabolic twin-breakwaters (10)	100-year return period (8)	Carbon footprint (4)
Hotspots & Nexus points (15)	Point absorbers and 5G-IoT (12)	Attenuators (9)	Wave excitation force (8)	Viscous heaving damping correction (4)
Electricity intermittency (15)	Energy management strategy (12)	Ocean energy harvesting (9)	Elastic cables (8)	Life cycle assessment (4)
Resource assessment through deep learning models (15)	Hydraulic motor (12)	Multi-body dynamics (9)	Free decay test (7)	Uncertain dynamics (3)
Shore power (15)	Stabilization (11)	National geodatabase (9)	Stokes wave theory (7)	Non-linear time-domain modeling (3)
Nearshore WECs (15)	Stochastic processes (11)	Random waves (9)	Deep q-network (7)	Variable inerter (3)
Climate change (15)	Stochastic control (11)	Power resource (9)	Heaving PA-WEC (7)	Adjustable draft system (3)
Coastal protection (15)	Hybrid WaveStar (11)	Elastic structure (9)	WECfarm (7)	Dynamic model (3)
Turbines (14)	Markov Chain Monte-Carlo (11)	Buoy geometry (9)	PACwave ocean conditions (7)	Surface gravity waves (3)
6 DOF PA-WEC (14)	Marine aquaculture (11)	Wave basic verification (9)	Adjustable natural period (7)	Semidefinite programming (3)
Multiple PTO units (14)	Design load (11)	Bioinspired WECs (9)	Wave-body interactions (7)	PTO force constraint (2)

TABLE 2.6
Operational robustness of the KCKS compared to the DCKS (in accordance with [59])

	KCKS					DCKS				
Verification yes (✓) or no (×)	**Average Centrality** (> 0.5)	**Average Silhouette** (> 0.85)	**Average TLS (top 30% items)**	**Recency (on a time scale)**	**Coverage (based on research frontiers)**	**Average Centrality** (> 0.5)	**Average Silhouette** (> 0.85)	**Average TLS (top 30% items)**	**Recency (on a time scale)**	**Coverage (based on research frontiers)**
Clustering	✓	✓	48.5	Till late 2024	–	× (< 0.2)	×	14.3	Till mid-2022	–
Hotspots	✓	–	86	Till late 2024	NA	×	–	57	Till 2021	NA
Frontiers	✓	NA	NA	Till late 2024	NA	×	NA	NA	Till 2023- end	NA
Knowledge structuring	NA	NA	NA	Till late 2024	High (compared to DCKS, 20% more frontiers)	NA	NA	NA	Till mid-2022	Low (compared to KCKS, 20% fewer frontiers)

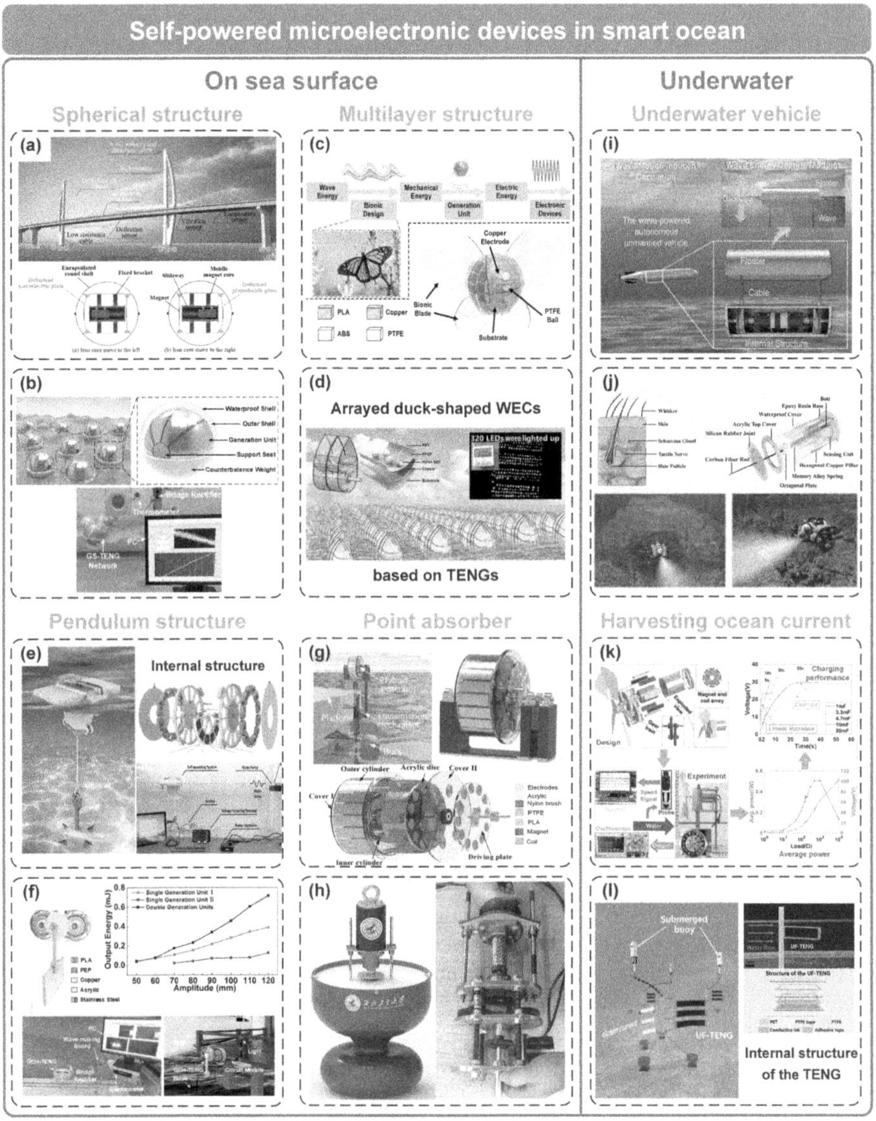

FIGURE 2.7 Self-powered and self-sensing micro-scale WECs and their smart sea applications [85](reproduced with the publisher's permission).

device was designed to improve its energy absorption bandwidth and durability, and the output of 400V could be achieved in 1.25 Hz waves, as demonstrated by lab-scale tests, making it viable for smart sea applications. The micro-WEC developed by Liu et al. [84] with a nodding duck-shaped floater design is shown in Figure 2.7(d), demonstrating a TENG with an enhanced energy capture bandwidth. Three covers of multi-rail TENGs were adapted to gain maximum space exploitation and to provide guide paths to the freely oscillating nylon balls for effective contact with triboelectric sheets inside the structure. The device achieved a maximum output of 4 Wm^{-3} and powered more than 300 LED bulbs, demonstrating a significant wave energy potential for self-powered smart sea gadgets.

Li et al. [86] developed a micro-scale WEC with an electromagnetic energy harvester comprising a pendulum-like structure, as shown in Figure 2.7(e), for monitoring water quality using a self-powered and self-sensing network of wireless sensor nodes. The device demonstrated an excellent energy capture with a maximum output of ~15mW, enough to run the sensors to monitor water quality. Similarly, Xu et al. [87] studied a pendulum-based TENG structure in a micro-WEC for structural health monitoring of offshore marine infrastructure, as shown in Figure 2.7(f). A set of two TENGs was developed for working separately under safe and extreme wave conditions, demonstrating a 2.3 times higher power performance compared to a single TENG set in a harsh marine environment.

Zhao et al. [88] investigated a hybrid micro heaving WEC comprising electromagnetic and triboelectric energy conversion modules embedded in a unique transmission system, as shown in Figure 2.7(g). Both the EMG and TENG modules in the device demonstrated an output of 2.5V and 2600V under wave conditions characterized by a height of 16 cm and a frequency of 0.8 Hz. The device facilitated an uninterrupted power supply to Bluetooth-enabled temperature and humidity monitoring sensors.

Ahmed et al. [29] conducted a study on optimizing floater geometry for enhancing energy capturing performance of the WECs utilizing optimal and suboptimal energy conversion modules in low-frequency waves, as shown in Figure 2.7(h). The WEC could achieve up to 17% and 22% increase in the hydrodynamic response and power absorption due to an optimally designed floater. Li et al. [89] demonstrated the design and performance assessment study of a self-powered autonomous underwater vehicle powered by a deployable WEC that could be retracted when the batteries are fully charged, as shown in Figure 2.7(i). The WEC-assisted vehicles demonstrated a significant potential for self-powered oceanographic and underwater research on aquatic life.

Liu et al. [90] enabled self-sensing detection of underwater obstruction to avoid collision of marine vessels and vehicles in marine environment, using a TENG comprising a whisker-resembling structure, as shown in Figure 2.7(j). The device comprised rods made of carbon fiber with a matrix structure and TENG modules to convert the external excitations into sensible electrical signals, enabling the sensing of external barriers in terms of displacement, direction, and intensity of load. Li et al. [91] studied self-powered undersea robots based on electromagnetic energy converters integrated in a compact circular structure and turbine motion amplified by the gears. Figure 2.7(k) shows a unique alternating EMG arrangement used to facilitate an abrupt increase in magnetic flux density, causing a significant increase in power performance of the WEC. Wang et al. [92] demonstrated a unique flag-shaped triboelectric energy converter for a micro-WEC enabled to harvest energy from ultra-low frequency currents, as shown in Figure 2.7(l). The lab-scale tests demonstrated a cumulative power output of 52.3 μW from six TENG modules under a flow velocity of ~0.5 ms^{-1}, enhancing its viability for self-powered smart sea applications, especially in low-frequency and low-speed ocean currents.

2.5.2 Macro-scale WECs

Macro- or large-scale WECs generally comprise floating buoys interacting with the waves, and the resultant forces are exploited through the PTO to generate electrical energy, with power outputs ranging between watts and kilowatts. The examples are WaveBob, Aquabuoy, PowerBuoy, and FO3 [93], [94]. They have a diameter smaller than the wave's length and absorb energy through oscillations about a reference position due to their interactions with waves. A single-body PA-WEC harvests energy through the reaction forces between a fixed structure and a wave-activated, floating body. For example, a single-body PA-WEC, as shown in Figure 2.8, comprising a spinning top-shaped buoy fixed through support with the PTO housed inside the buoy was investigated by Azam et al. [36]. The proposed PA-WEC was demonstrated for applications in aquaculture cage for sustainable fishing and oceanography. A two-body PA-WEC exploits the reaction forces between a submerged body and a floating body. The power performance of a two-body PA-WEC can be significantly greater than that of a single-body PA-WEC with the same floating body [95]. Research on PA-WECs is emerging, and various studies have enhanced their applicability and attainability. The recent projects have frequently focused on PA-WEC

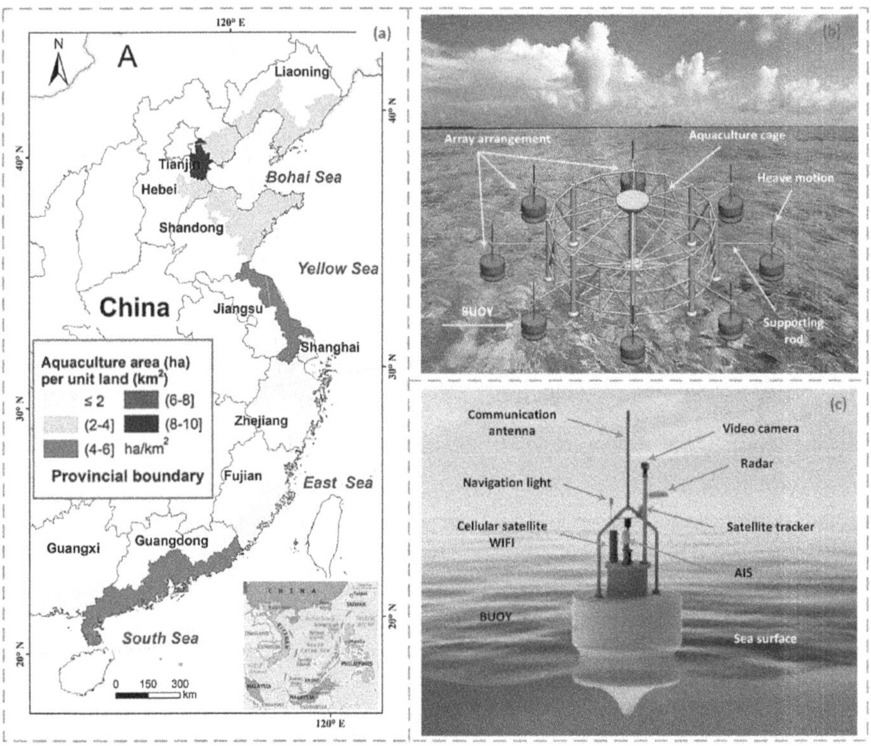

FIGURE 2.8 (a) Coastal areas of China, (b) arrayed floating buoys with aquaculture cage for sustainable fishing, (c) application in the oceanography [36] (adapted with the permission of the publishers).

design, control strategies, PTO power performance [96], buoy geometry optimization, moorings [97], feasible applications, arrayed installations [98], open-sea tests [99], lab-scale testing, wave-to-wire modeling, and AI-based technologies to assist their commercialization. A few successful macro-PA-WECs are summarized below.

2.5.2.1 The OPT PowerBuoy

The PB3 PowerBuoy, developed by Ocean Power Technologies [100], uses its two-body PA-WEC to provide an uninterruptible power supply to its navigation sensors. It uses a direct-drive generator to charge its battery, is moored at the seabed, and can operate in ocean depths between 20 and 3000m [101]. It can continuously supply energy to the equipment placed on the seabed and transfer real-time data to onshore facilities.

2.5.2.2 The CorPower WEC

The CorPower WEC [102] mimics the human heart's pumping mechanism to build up pressure used to harvest five times more energy per ton of the device compared to the previous designs. The CorPower facilitates robust and cheaper energy harvesting using a heaving buoy that exploits the waves' heave and surge motions. The buoy is moored at the seabed through the PTO, and the wave-buoy interactions are controlled to maximize power absorption. The CorPower's C4 array, the world's first grid-connected wave farm, is planned to be installed off the coast of Aguçadoura, Portugal [103]. Each C4 has a rated capacity of 300 kW, making it the most compact PA-WEC in terms of its power output [104].

2.5.2.3 The Wave Star WEC

The commercial WaveStar WEC [105] employs hydraulic PTO absorbing energy through the arrays of hemispherical-bottomed buoys. A pressure difference generated across the hydraulic cylinders is used to run the hydraulic motor. The Wavestar C6–600 prototypes and Hanstholm prototypes have been installed on the western coasts of Denmark. The hydraulic PTO has been tested to attain a maximum of 60% overall efficiency [106]. The Wavestar provides three major advantages: (i) it uses floater arrays causing greater output power, (ii) it facilitates convenient storm protection, and (iii) the platform and PTO are installed above water, and only buoys are floating, offering a reliable design [107].

2.5.2.4 The Azura WEC

A demonstration-level Azura WEC developed and deployed by Azura™ Ocean Wave Technology exploits both heave and surge motions (the relative motion between the floating buoy and the hull) to produce electricity from waves. Two prototypes, each with a rated capacity of 20 kW, were deployed for testing in Oregon and New Zealand, subsequently, tested by University of Hawaii for 1.5 years. The prototype was successful with an overall efficiency greater than 50% and an operational uptime of ~98% in waves higher than 7.5m.

2.6 CRITICAL CHALLENGES AND FUTURE RESEARCH DIRECTIONS

This section highlights the critical challenges that underscore the intricate and interdisciplinary nature of advancing AI-assisted WECs. Overcoming these obstacles will be crucial for their successful commercialization and broader implementation. Efforts are required to achieve a comparable levelized cost of energy (LCOE) for WECs in comparison to other renewables, such as wind and solar energy systems [108, 109]. The integration of AI-assisted frameworks in recent WEC systems encounters significant challenges, including high production costs, reliability concerns, frequent maintenance demands, elevated energy costs, survivability issues, energy efficiency limitations, the operational lifespan, and ecological risks [110]. The availability of comprehensive wave climate data necessitates precise assessment, with data collection required from satellites, offshore buoys, and remote sources [111]. The performance of WECs is significantly influenced by the design of the floater, as well as its response and optimal interaction with the PTO system [112]. Berenjkoob et al. [113] suggested considering wave-structure interactions and viscous drag effects when investigating the influence of changing the buoy shapes.

Future studies are strongly encouraged to focus on multi-degree-of-freedom WECs for enhanced energy extraction from heave, surge, and pitch motions simultaneously. Additionally, exploring new marine sites and conducting feasibility studies are essential to accurately assess the real-time viability of WEC devices in marine applications [114]. Conventional linearized models have proven to be inaccurate when accounting for non-linearities caused by viscous drag in the study of larger WECs, where significant viscous forces are involved. High-fidelity simulation models that integrate numerical wave tanks with non-linear PTO constraints need to be thoroughly investigated and experimentally validated. This includes addressing complex non-linear phenomena such as wake effects, turbulence, and flow separation [115]. Developing a WEC system that can vary its resonance frequency by adapting to an optimal geometry, inertia, mass, and draft in real-time to optimize energy capture bandwidth according to changing sea conditions for maximum power absorption is a challenging task [116]. For commercialization and technological advancement, comparing different types of WECs across various sea states is essential to identify the optimal WEC for each condition. Achieving long-term survivability and reliability in open-sea environments requires determining the most effective harvesting technique.

For WECs tested in small-scale wave tanks, prototype testing in an open-sea environment is essential to accurately predict their power performance, service life, operational characteristics, and maintenance requirements [117]. Investigations on the structural reliability of WECs using failure mode and effects analyses are rare. Testing WEC systems for operation along shorelines or in breakwaters is complex and costly even in laboratories [118]. The testing method is significantly influenced by the researcher's objectives, harvesting approach, available funding, and infrastructure. Uncertainties in the operational and performance characteristics of WECs increase significantly when transitioning from simulated wave conditions to real-time sea conditions or when scaling up from a model to a full-scale prototype.

Due to the harsh marine environment and high salinity, WECs are prone to damage and corrosion, leading to increased maintenance requirements. Investigating the environmental impact of different WEC types can be challenging and inconvenient. During storms and hurricanes, their operation, performance, and survival become major concerns. Therefore, essential maintenance and protective strategies should be incorporated during the design phase, which will likely increase the overall lifecycle costs. Additionally, the installation and operation of WECs should not negatively impact the marine ecosystem and underwater life. During installation, noise pollution, entrapment, and collisions caused by external activities can disrupt marine life and their habitats. Issues related to aesthetics, transmission lines, and moorings should also be addressed in planning. It is essential to develop strategies that ensure long-term harmony between natural environments and human activities before commencing an underwater project.

REFERENCES

[1] A. Azam, A. Ahmed, M. Yi, Z. Zhang, X. Tan, A. Ali, and N. Li, "A self-stabilizing point absorber wave energy converter with a top-shaped buoy and non-linear power take-off for oceanographic applications," *Ocean Eng.*, vol. 288, p. 116018, 2023.

[2] H. Li, J. Wu, X. Shi, L. Kong, W. Kong, Z. Zhang, Y. Pan, D. Luo, and J. Yan, "A self-powered smart wave energy converter for sustainable sea," *Mech. Syst. Signal Process.*, vol. 220, p. 111641, 2024.

[3] T. Ogata, Y. Washio, H. Osawa, Y. Tsuritani, S. Yamashita, and Y. Nagata, "The open sea tests of the offshore floating type wave power device 'Mighty Whale': Performance of the prototype," in *International Conference on Offshore Mechanics and Arctic Engineering*, 2002, pp. 517–524.

[4] H. Ohneda, S. Igarashi, O. Shinbo, S. Sekihara, K. Suzuki, H. Kubota, and H. Morita, "Construction procedure of a wave power extracting caisson breakwater," in *Proceedings of 3rd Symposium on Ocean Energy Utilization*, 1991, pp. 171–179.

[5] A. Ahmed, A. Azam, Y. Wang, X. Tan, M. Yi, and Z. Zhang, "On the S-shaped floaters for a Wavestar-like wave energy converter with an I-shaped mechanical power take-off," *Energy Convers. Manag. X*, p. 100387, 2023.

[6] C. Eskilsson, J. Palm, J. P. Kofoed, and E. Friis-Madsen, "CFD study of the overtopping discharge of the Wave Dragon wave energy converter," *Renew. Energies Offshore*, pp. 287–294, 2015.

[7] ScienceDirect, "Advanced search," Elsevier. Accessed: Dec. 06, 2022. [Online]. Available: https://www.sciencedirect.com/search

[8] J. Hu, B. Zhou, C. Vogel, P. Liu, R. Willden, K. Sun, J. Zang, J. Geng, P. Jin, L. Cui, and B. Jiang, "Optimal design and performance analysis of a hybrid system combing a floating wind platform and wave energy converters," *Appl. Energy*, vol. 269, p. 114998, 2020.

[9] D. Younesian and M.-R. Alam, "Multi-stable mechanisms for high-efficiency and broadband ocean wave energy harvesting," *Appl. Energy*, vol. 197, pp. 292–302, 2017.

[10] International renewable energy agency, "Wave Energy: Technology brief." Accessed: Sep. 23, 2022. [Online]. Available: https://www.irena.org

[11] S. Zheng, Y. Zhang, and G. Iglesias, "Concept and performance of a novel wave energy converter: Variable Aperture Point-Absorber (VAPA)," *Renew. Energy*, vol. 153, pp. 681–700, 2020.

[12] P. Contestabile, C. Iuppa, E. Di Lauro, L. Cavallaro, T. L. Andersen, and D. Vicinanza, "Wave loadings acting on innovative rubble mound breakwater for overtopping wave energy conversion," *Coast. Eng.*, vol. 122, pp. 60–74, 2017.

[13] J. C. C. Henriques, L. M. C. Gato, A. F. de O. Falcao, E. Robles, and F.-X. Faÿ, "Latching control of a floating oscillating-water-column wave energy converter," *Renew. Energy*, vol. 90, pp. 229–241, 2016.

[14] H. Hu, W. Xue, P. Jiang, and Y. Li, "Bibliometric analysis for ocean renewable energy: An comprehensive review for hotspots, frontiers, and emerging trends," *Renew. Sustain. Energy Rev.*, vol. 167, p. 112739, 2022.

[15] S.-H. Yang, J. W. Ringsberg, E. Johnson, and Z. Hu, "Biofouling on mooring lines and power cables used in wave energy converter systems—Analysis of fatigue life and energy performance," *Appl. Ocean Res.*, vol. 65, pp. 166–177, 2017.
[16] C. W. Zheng and C. Y. Li, "Variation of the wave energy and significant wave height in the China Sea and adjacent waters," *Renew. Sustain. Energy Rev.*, vol. 43, pp. 381–387, 2015.
[17] C. Rodriguez-Delgado and R. J. Bergillos, "Wave energy assessment under climate change through artificial intelligence," *Sci. Total Environ.*, vol. 760, p. 144039, 2021.
[18] Z. Lv, N. Wang, R. Lou, Y. Tian, and M. Guizani, "Towards carbon Neutrality: Prediction of wave energy based on improved GRU in Maritime transportation," *Appl. Energy*, vol. 331, p. 120394, 2023.
[19] A. Shadmani, M. R. Nikoo, A. H. Gandomi, and M. Chen, "An optimization approach for geometry design of multi-axis wave energy converter," *Energy*, vol. 301, p. 131714, 2024.
[20] L. Yang, J. Huang, J. Mi, M. Hajj, G. Bacelli, and L. Zuo, "Control-inspired design and power optimization of an active mechanical motion rectifier based power takeoff for wave energy converters," *Renew. Energy*, vol. 228, p. 120614, 2024.
[21] A. P. Zhao, S. Li, D. Xie, P. J. H. Hu, C. Wu, F. X. Fei, T. T. Li, Y. Xiang, C. Gu, and Z. Cao, "Extreme events threat water-energy-carbon nexus through cascading effects," *Next Energy*, vol. 5, p. 100151, 2024.
[22] S. Bhattacharya and B. K. Sachdev, "Impact of scientific innovation in coastal resource management: Fostering blue economic development, environmental mitigation, and pollution control," in *Scientific Innovations for Coastal Resource Management*, IGI Global, 2024, pp. 1–32.
[23] C. Gu and H. Li, "Review on deep learning research and applications in wind and wave energy," *Energies*, vol. 15, no. 4, p. 1510, 2022.
[24] C. Dong, G. Xu, G. Han, B. J. Bethel, W. Xie, and S. Zhou, "Recent developments in artificial intelligence in oceanography," *OLAR*, vol. 2022, p. 26, 2022. Article ID: 9870950.
[25] M. Ali, R. Prasad, Y. Xiang, A. Sankaran, R.C. Deo, F. Xiao, and S. Zhu, "Advanced extreme learning machines vs. deep learning models for peak wave energy period forecasting: A case study in Queensland, Australia," *Renew. Energy*, vol. 177, pp. 1031–1044, 2021.
[26] R. Savitha and A. Al Mamun, "Regional ocean wave height prediction using sequential learning neural networks," *Ocean Eng.*, vol. 129, pp. 605–612, 2017.
[27] J. Wang, L. Aouf, and S. Badulin, "Retrieval of wave period from altimetry: Deep learning accounting for random wave field dynamics," *Remote Sens. Environ.*, vol. 265, p. 112629, 2021.
[28] S. Jin, R. J. Patton, and B. Guo, "Enhancement of wave energy absorption efficiency via geometry and power take-off damping tuning," *Energy*, vol. 169, pp. 819–832, 2019.
[29] A. Ahmed, Y. Wang, A. Azam, and Z. Zhang, "Design and analysis of the bulbous-bottomed oscillating resonant buoys for an optimal point absorber wave energy converter," *Ocean Eng.*, vol. 263, p. 112443, 2022.
[30] P. S. Kumar, K. Dongeun, and L. Yeonbin, "Design optimization of asymmetric wave energy converter using artificial neural network model," *Int. J. Nav. Archit. Ocean Eng.*, vol. 15, pp. 1–12, 2023.
[31] A. Ahmed, A. Azam, Z. Zhang, M. Yi, T. Aslam, S. A. Mugheri, D. Hao, M. Abdelrahman, and A. M. Tairab, "An isomorphic Froude scaling approach to bulbous-bottomed buoys in wave energy converters for smart floating cities," *Appl. Ocean Res.*, vol. 156, p. 104485, 2025.
[32] A. Ahmed, Y. Wang, A. Azam, N. Li, C. Jia, and Z. Zhang, "Design of an S-shaped point-absorber wave energy converter with a non-linear PTO to power the satellite-respondent buoys in the East China Sea," *Ocean Eng.*, vol. 275, p. 114162, 2023.
[33] M. S. Nazir, Z. M. Ali, M. Bilal, H. M. Sohail, and H. M. N. Iqbal, "Environmental impacts and risk factors of renewable energy paradigm—a review," *Environ. Sci. Pollut. Res.*, vol. 27, pp. 33516–33526, 2020.

[34] I. Galparsoro, M. Korta, I. Subirana, Á. Borja, I. Menchaca, O. Solaun, I. Muxika, G. Iglesias, and J. Bald, "A new framework and tool for ecological risk assessment of wave energy converters projects," *Renew. Sustain. Energy Rev.*, vol. 151, p. 111539, 2021.

[35] S. S. Kulkarni and D. J. Edwards, "A bibliometric review on the implications of renewable offshore marine energy development on marine species," *Aquac. Fish.*, vol. 7, no. 2, pp. 211–222, 2022.

[36] A. Azam, A. Ahmed, H. Li, A. M. Tairab, C. Jia, N. Li, and Z. Zhang, "Design and analysis of the optimal spinning top-shaped buoy for wave energy harvesting in low energy density seas for sustainable marine aquaculture," *Ocean Eng.*, vol. 255, p. 111434, 2022.

[37] B. Yang, S. Wu, H. Zhang, B. Liu, H. Shu, J. Shan, Y. Ren, and W. Yao, "Wave energy converter array layout optimization: A critical and comprehensive overview," *Renew. Sustain. Energy Rev.*, vol. 167, p. 112668, 2022.

[38] K. Zhu, H. Shi, M. Han, and F. Cao, "Layout study of wave energy converter arrays by an artificial neural network and adaptive genetic algorithm," *Ocean Eng.*, vol. 260, p. 112072, 2022.

[39] M. G. J. C. M. Kamarlouei, J. F. Gaspar, M. Calvário, T. S. Hallak, M. J. Mendes, F. Thiebaut, and C. G. Soares, "Experimental analysis of wave energy converters concentrically attached on a floating offshore platform," *Renew. Energy*, vol. 152, pp. 1171–1185, 2020.

[40] A. Azam, A. Ahmed, A. Qammar, A. N. Shah, Z. Zhang, Z. Zhang, T. Aslam, S.A. Mugheri, M. Yi, and X. Tan, "From model-scale to full-scale: Optimizing buoy geometry for enhanced energy conversion in PA-WECs," *Ocean Eng.*, vol. 332, p. 121458, 2025.

[41] E. Anderlini, D. I. M. Forehand, E. Bannon, and M. Abusara, "Reactive control of a wave energy converter using artificial neural networks," *Int. J. Mar. energy*, vol. 19, pp. 207–220, 2017.

[42] AlHadrami, A.A., Chala, G.T., and Alkalbani, A.M. (2026). Wave energy converters and AI. In: Sulaiman, S.A. (eds) *Artificial Intelligence in Sustainable Energy and Environmental Management*. Springer, Singapore. https://doi.org/10.1007/978-981-95-6148-3_3.

[43] A. Shadmani, M. R. Nikoo, A. H. Gandomi, R.-Q. Wang, and B. Golparvar, "A review of machine learning and deep learning applications in wave energy forecasting and WEC optimization," *Energy Strateg. Rev.*, vol. 49, p. 101180, 2023.

[44] E. Hultman and D. Salar, "A robotized 6-DOF dry test rig for wave power," *Sustain. Energy Technol. Assessments*, vol. 59, p. 103393, 2023.

[45] N.-Z. Chen, Z. Zhao, and L. Lin, "A hybrid deep learning method for AE source localization for heterostructure of wind turbine blades," *Mar. Struct.*, vol. 94, p. 103562, 2024.

[46] J. Li, J. Lian, Y. Guo, and H. Wang, "Concept design and floating installation method study of multi-bucket foundation floating platform for offshore wind turbines," *Mar. Struct.*, vol. 93, p. 103541, 2024.

[47] P. Guo, H. Ling, Y. Zhang, and T. Di, "Modelling of foundation damping in time-domain analysis of monopile supported offshore wind turbine structures," *Mar. Struct.*, vol. 98, p. 103672, 2024.

[48] L. Li, Y. Gao, D. Z. Ning, and Z. M. Yuan, "Development of a constraint non-causal wave energy control algorithm based on artificial intelligence," *Renew. Sustain. Energy Rev.*, vol. 138, p. 110519, 2021.

[49] A. Maria-Arenas, A. J. Garrido, E. Rusu, and I. Garrido, "Control strategies applied to wave energy converters: State of the art," *Energies*, vol. 12, no. 16, p. 3115, 2019.

[50] S. Sarkar, V. Gundecha, A. Shmakov, S. Ghorbanpour, A. R. Babu, P. Faraboschi, M. Cocho, A. Pichard, and J. Fievez, "Multi-agent reinforcement learning controller to maximize energy efficiency for multi-generator industrial wave energy converter," in *Proceedings of the AAAI Conference on Artificial Intelligence*, 2022, pp. 12135–12144.

[51] S. Zou, X. Zhou, I. Khan, W. W. Weaver, and S. Rahman, "Optimization of the electricity generation of a wave energy converter using deep reinforcement learning," *Ocean Eng.*, vol. 244, p. 110363, 2022.

[52] A. Kassayeva, A. Balekova, and K. Yerzhanov, "Converting the energy of sea waves into electrical energy," *Environ. Prog. Sustain. Energy*, vol. 43, no. 3, p. e14342, 2024.

[53] M. Rosati, H. A. Said, and J. V. Ringwood, "Wave-to-wire control of an oscillating water column wave energy system equipped with a wells turbine," in *Proceedings of the 15th European Wave and Tidal Energy Conference*, vol. 15, 2023. https://doi.org/10.36688/ewtec-2023-309.

[54] A. George and I. H. Cho, "Implementing nonlinear viscous damping of a heave disk with specific porosities using machine learning models in time domain," *Ocean Eng.*, vol. 297, p. 117142, 2024.

[55] S. Xu, S. Wang, and C. G. Soares, "Prediction of mooring tensions of a wave energy converter considering the effects of nonlinear axial stiffness by a deep learning neural network," *Ocean Eng.*, vol. 305, p. 117810, 2024.

[56] Y. Wang, A. Ahmed, A. Azam, D. Bing, Z. Shan, Z. Zhang, M. K. Tariq, J. Sultana, R. T. Mushtaq, A. Mehboob, and C. Xiaohu, "Applications of additive manufacturing (AM) in sustainable energy generation and battle against COVID-19 pandemic: The knowledge evolution of 3D printing," *J. Manuf. Syst.*, vol. 60, pp. 709–733, 2021.

[57] A. Azam, A. Ahmed, H. Wang, Y. Wang, and Z. Zhang, "Knowledge structure and research progress in wind power generation (WPG) from 2005 to 2020 using CiteSpace based scientometric analysis," *J. Clean. Prod.*, vol. 295, p. 126496, 2021.

[58] A. Azam, A. Ahmed, M. S. Kamran, L. Hai, Z. Zhang, and A. Ali, "Knowledge structuring for enhancing mechanical energy harvesting (MEH): An in-depth review from 2000 to 2020 using CiteSpace," *Renew. Sustain. Energy Rev.*, vol. 150, p. 111460, 2021.

[59] A. Ahmed, A. Azam, M. M. Aslam Bhutta, F. A. Khan, R. Aslam, and Z. Tahir, "Discovering the technology evolution pathways for 3D printing (3DP) using bibliometric investigation and emerging applications of 3DP during COVID-19," *Clean. Environ. Syst.*, vol. 3, p. 100042, 2021.

[60] F. de O. Antonio, "Wave energy utilization: A review of the technologies," *Renew. Sustain. Energy Rev.*, vol. 14, no. 3, pp. 899–918, 2010.

[61] J. Falnes, "*Ocean Waves and Oscillating Systems: Linear Interactions Including Wave-energy Extraction*," p. 275, Mar. 2002. Cambridge University Press. ISBN: 9780511754630. https://doi.org/10.1017/CBO9780511754630.

[62] B. Drew, A. R. Plummer, and M. N. Sahinkaya, "A review of wave energy converter technology," 2009, Sage Publications Sage UK: London, England.

[63] J. Falnes, "A review of wave-energy extraction," *Mar. Struct.*, vol. 20, no. 4, pp. 185–201, 2007.

[64] K. Budar and J. Falnes, "A resonant point absorber of ocean-wave power," *Nature*, vol. 256, no. 5517, pp. 478–479, 1975.

[65] A. Babarit, J. Hals, M. J. Muliawan, A. Kurniawan, T. Moan, and J. Krokstad, "Numerical benchmarking study of a selection of wave energy converters," *Renew. Energy*, vol. 41, pp. 44–63, 2012.

[66] A. Clément, P. McCullen, A. Falcão, A. Fiorentino, F. Gardner, K. Hammarlund, G. Lemonis, T. Lewis, K. Nielsen, S. Petroncini, and M. T. Pontes, "Wave energy in Europe: Current status and perspectives," *Renew. Sustain. Energy Rev.*, vol. 6, no. 5, pp. 405–431, 2002.

[67] A. Babarit and A. H. Clément, "Optimal latching control of a wave energy device in regular and irregular waves," *Appl. Ocean Res.*, vol. 28, no. 2, pp. 77–91, 2006.

[68] D. V. Evans, "A theory for wave-power absorption by oscillating bodies," *J. Fluid Mech.*, vol. 77, no. 1, pp. 1–25, 1976.

[69] M. Vantorre, R. Banasiak, and R. Verhoeven, "Modelling of hydraulic performance and wave energy extraction by a point absorber in heave," *Appl. Ocean Res.*, vol. 26, no. 1–2, pp. 61–72, 2004.
[70] Y. Li and Y.-H. Yu, "A synthesis of numerical methods for modeling wave energy converter-point absorbers," *Renew. Sustain. Energy Rev.*, vol. 16, no. 6, pp. 4352–4364, 2012.
[71] S. J. Beatty, M. Hall, B. J. Buckham, P. Wild, and B. Bocking, "Experimental and numerical comparisons of self-reacting point absorber wave energy converters in regular waves," *Ocean Eng.*, vol. 104, pp. 370–386, 2015.
[72] J. V. Ringwood, G. Bacelli, and F. Fusco, "Energy-maximizing control of wave-energy converters: The development of control system technology to optimize their operation," *IEEE Control Syst. Mag.*, vol. 34, no. 5, pp. 30–55, 2014.
[73] J. Hals, J. Falnes, and T. Moan, "Constrained optimal control of a heaving buoy wave-energy converter," *J. Offshore Mech. Arct. Eng.*, vol. 133, no. 1, pp. 15, 2011. Article ID: 011401.
[74] I. López, J. Andreu, S. Ceballos, I. M. De Alegría, and I. Kortabarria, "Review of wave energy technologies and the necessary power-equipment," *Renew. Sustain. Energy Rev.*, vol. 27, pp. 413–434, 2013.
[75] R. Taghipour, T. Perez, and T. Moan, "Hybrid frequency–time domain models for dynamic response analysis of marine structures," *Ocean Eng.*, vol. 35, no. 7, pp. 685–705, 2008.
[76] S. Bozzi, A. M. Miquel, A. Antonini, G. Passoni, and R. Archetti, "Modeling of a point absorber for energy conversion in Italian seas," *Energies*, vol. 6, no. 6, pp. 3033–3051, 2013.
[77] M. Shadman, S. F. Estefen, C. A. Rodriguez, and I. C. M. Nogueira, "A geometrical optimization method applied to a heaving point absorber wave energy converter," *Renew. Energy*, vol. 115, pp. 533–546, 2018.
[78] A. Babarit, G. Duclos, and A. H. Clément, "Comparison of latching control strategies for a heaving wave energy device in random sea," *Appl. Ocean Res.*, vol. 26, no. 5, pp. 227–238, 2004.
[79] F. de O. António, "Modelling and control of oscillating-body wave energy converters with hydraulic power take-off and gas accumulator," *Ocean Eng.*, vol. 34, no. 14–15, pp. 2021–2032, 2007.
[80] L. Kong, Z. Fang, T. Zhang, Z. Zhang, Y. Pan, D. Hao, J. Chen, and L. Qi, "A self-powered and self-sensing lower-limb system for smart healthcare," *Adv. Energy Mater.*, vol. 13, no. 31, p. 2301254, 2023.
[81] L. Qi, H. Li, X. Wu, Z. Zhang, W. Duan, and M. Yi, "A hybrid piezoelectric-electromagnetic wave energy harvester based on capsule structure for self-powered applications in sea-crossing bridges," *Renew. Energy*, vol. 178, pp. 1223–1235, 2021.
[82] Q. Gao, Y. Xu, X. Yu, Z. Jing, T. Cheng, and Z. L. Wang, "Gyroscope-structured triboelectric nanogenerator for harvesting multidirectional ocean wave energy," *ACS Nano*, vol. 16, no. 4, pp. 6781–6788, 2022.
[83] X. Wang, Q. Gao, M. Zhu, J. Wang, J. Zhu, H. Zhao, Z. L. Wang, and T. Cheng, "Bioinspired butterfly wings triboelectric nanogenerator with drag amplification for multidirectional underwater-wave energy harvesting," *Appl. Energy*, vol. 323, p. 119648, 2022.
[84] L. Liu, X. Yang, L. Zhao, H. Hong, H. Cui, J. Duan, Q. Yang, and Q. Tang, "Nodding duck structure multi-track directional freestanding triboelectric nanogenerator toward low-frequency ocean wave energy harvesting," *ACS Nano*, vol. 15, no. 6, pp. 9412–9421, 2021.
[85] L. Qi, L. Kong, Y. Wang, J. Song, A. Azam, Z. Zhang, and J. Yan, "Recent progress in application-oriented self-powered microelectronics," *Adv. Energy Mater.*, vol. 13, no. 47, p. 2302699, 2023.

[86] M. Li, Y. Zhang, K. Li, Y. Zhang, K. Xu, X. Liu, S. Zhong, and J. Cao, "Self-powered wireless sensor system for water monitoring based on low-frequency electromagnetic-pendulum energy harvester," *Energy*, vol. 251, p. 123883, 2022.
[87] Y. Xu, W. Yang, X. Lu, Y. Yang, J. Li, J. Wen, T. Cheng, and Z. L. Wang, "Triboelectric nanogenerator for ocean wave graded energy harvesting and condition monitoring," *ACS Nano*, vol. 15, no. 10, pp. 16368–16375, 2021.
[88] B. Zhao, Z. Li, X. Liao, L. Qiao, Y. Li, S. Dong, Z. Zhang, and B. Zhang, "A heaving point absorber-based ocean wave energy convertor hybridizing a multilayered soft-brush cylindrical triboelectric generator and an electromagnetic generator," *Nano Energy*, vol. 89, p. 106381, 2021.
[89] H. Li, X. Wu, Z. Zhang, X. Tan, Y. Pan, C. Dai, D. Luo, A. Ahmed, and Y. Xu, "An extended-range wave-powered autonomous underwater vehicle applied to underwater wireless sensor networks," *Iscience*, vol. 25, no. 8, p. 104738, 2022.
[90] J. Liu, P. Xu, J. Zheng, X. Liu, X. Wang, S. Wang, T. Guan, G. Xie, and M. Xu, "Whisker-inspired and self-powered triboelectric sensor for underwater obstacle detection and collision avoidance," *Nano Energy*, vol. 101, p. 107633, 2022.
[91] Z. Li, X. Jiang, P. Yin, L. Tang, H. Wu, Y. Peng, J. Luo, S. Xie, H. Pu, and D. Wang, "Towards self-powered technique in underwater robots via a high-efficiency electromagnetic transducer with circularly abrupt magnetic flux density change," *Appl. Energy*, vol. 302, p. 117569, 2021.
[92] Y. Wang, X. Liu, T. Chen, H. Wang, C. Zhu, H. Yu, L. Song, X. Pan, J. Mi, C. Lee, and M. Xu, "An underwater flag-like triboelectric nanogenerator for harvesting ocean current energy under extremely low velocity condition," *Nano Energy*, vol. 90, p. 106503, 2021.
[93] C. Liang, J. Ai, and L. Zuo, "Design, fabrication, simulation and testing of an ocean wave energy converter with mechanical motion rectifier," *Ocean Eng.*, vol. 136, pp. 190–200, 2017.
[94] C. Liu, "A tunable resonant oscillating water column wave energy converter," *Ocean Eng.*, vol. 116, pp. 82–89, 2016.
[95] C. Liang and L. Zuo, "On the dynamics and design of a two-body wave energy converter," in *Journal of Physics: Conference Series*, IOP Publishing, 2016, p. 12074.
[96] Y. Zhang, Y. Zhao, W. Sun, and J. Li, "Ocean wave energy converters: Technical principle, device realization, and performance evaluation," *Renew. Sustain. Energy Rev.*, vol. 141, p. 110764, 2021.
[97] M. N. Berenjkoob, M. Ghiasi, and C. G. Soares, "Performance of two types of mooring systems in the heave motion of a two-body floating wave energy converter," *J. Mar. Sci. Appl.*, vol. 18, no. 1, pp. 38–47, 2019.
[98] P. Sun, S. Hu, H. He, S. Zheng, H. Chen, S. Yang, and Z. Ji., "Structural optimization on the oscillating-array-buoys for energy-capturing enhancement of a novel floating wave energy converter system," *Energy Convers. Manag.*, vol. 228, p. 113693, 2021.
[99] S. Qiu, K. Liu, D. Wang, J. Ye, and F. Liang, "A comprehensive review of ocean wave energy research and development in China," *Renew. Sustain. Energy Rev.*, vol. 113, p. 109271, 2019.
[100] C.-N. Wang, T.-D. T. Pham, and D.-B. Nguyen, "A study of appropriate wave energy technology for sustainable development in Australia," *J. Eng. Res.*, vol. 13, issue 3, pp. 2589–2602, 2024.
[101] Ocean Power Technologies, "The OPT PB3 PowerBuoy." Accessed: Sep. 24, 2022. [Online]. Available: https://oceanpowertechnologies.com
[102] Patrik Möller, "CorPower C4 WEC," CorPower Ocean. Accessed: Jan. 11, 2023. [Online]. Available: https://corpowerocean.com
[103] X. Shao, J. W. Ringsberg, H.-D. Yao, U. R. S. L. Gowda, H. N. Khedkar, and J. H. Todalshaug, "Hydrodynamic interactions and enhanced energy harnessing amongst many WEC units in large-size wave parks," *J. Mar. Sci. Eng.*, vol. 12, no. 5, p. 730, 2024.

[104] R. Satymov, D. Bogdanov, M. Dadashi, G. Lavidas, and C. Breyer, "Techno-economic assessment of global and regional wave energy resource potentials and profiles in hourly resolution," *Appl. Energy*, vol. 364, p. 123119, 2024.

[105] Z. Luan, B. Chen, R. Jin, G. He, H. Ghassemi, and P. Jing, "Validation of a numerical wave tank based on overset mesh for the wavestar-like wave energy converter in the South China Sea," *Energy*, vol. 290, p. 130192, 2024.

[106] Niels and Keld Hansen, "Wavestar prototype at Roshage," Wave Star Unlimited Clean Energy. Accessed: Aug. 15, 2022. [Online]. Available: https://wavestarenergy.com

[107] Z. Zhang, X. Chen, H. Wu, W. Liu, and L. Cui, "Numerical study of a novel hybrid system with the Wavestar wave energy converter array and a SPIC semi-submersible floating platform," *J. Clean. Prod.,* vol. 407, p. 137178, 2023.

[108] W. Azam, I. Khan, and S. A. Ali, "Alternative energy and natural resources in determining environmental sustainability: a look at the role of government final consumption expenditures in France," *Environ. Sci. Pollut. Res.*, vol. 30, no. 1, pp. 1949–1965, 2023.

[109] A. Usman, S. Ullah, I. Ozturk, M. Z. Chishti, and S. M. Zafar, "Analysis of asymmetries in the nexus among clean energy and environmental quality in Pakistan," *Environ. Sci. Pollut. Res.*, vol. 27, pp. 20736–20747, 2020.

[110] A. Bielecki, S. Ernst, W. Skrodzka, and I. Wojnicki, "The externalities of energy production in the context of development of clean energy generation," *Environ. Sci. Pollut. Res.*, vol. 27, pp. 11506–11530, 2020.

[111] A. Garcia-Teruel and D. I. M. Forehand, "A review of geometry optimisation of wave energy converters," *Renew. Sustain. Energy Rev.*, vol. 139, p. 110593, 2021.

[112] L. Rusu and F. Onea, "The performance of some state-of-the-art wave energy converters in locations with the worldwide highest wave power," *Renew. Sustain. Energy Rev.*, vol. 75, pp. 1348–1362, 2017.

[113] M. Nazari Berenjkoob, M. Ghiasi, and C. Guedes Soares, "On the improved design of the buoy geometry on a two-body wave energy converter model," *J. Renew. Sustain. Energy*, vol. 11, no. 5, p. 54702, 2019.

[114] A. Azam, A. Ahmed, M. Yi, Z. Zhang, Z. Zhang, T. Aslam, S. A. Mugheri, M. Abdelrahman, A. Ali, and L. Qi, "Wave energy evolution: Knowledge structure, advancements, challenges and future opportunities," *Renew. Sustain. Energy Rev.*, vol. 205, p. 114880, 2024.

[115] A. Shadmani, M. R. Nikoo, A. H. Gandomi, M. Chen, and R. Nazari, "Advancements in optimizing wave energy converter geometry utilizing metaheuristic algorithms," *Renew. Sustain. Energy Rev.*, vol. 197, p. 114398, 2024.

[116] B. W. Schubert, B. W. Schubert, W. S. P. Robertson, B. S. Cazzolato, N. Y. Sergiienko, and M. H, Ghayesh., "Nonlinear stiffness enhancement of submerged wave energy device in high fidelity model," *Ocean Eng.*,vol, 254, pp. 111295, 2022.

[117] H. Luo, S. Cao, and V. L. Lu, "The techno-economic feasibility of a coastal zero-energy hotel building supported by the hybrid wind–wave energy system," *Sustain. Energy, Grids Networks*, vol. 30, p. 100650, 2022.

[118] G. He, Z. Luan, W. Zhang, R. He, C. Liu, K. Yang, C. Yang, P. Jing, and Z. Zhang, "Review on research approaches for multi-point absorber wave energy converters," *Renew. Energy*, vol. 218, p. 119237, 2023.

3 IoT-Driven Smart Port Logistics for Renewable Energy Integration in Maritime Supply Chains

Ish Kumar and Vinay Maitri

3.1 INTRODUCTION

Maritime transport continues to underpin international trade, with sea carriage dominating global freight volumes. Ports therefore sit at the centre of logistics performance, regional competitiveness, and environmental externalities, which has sharpened the policy focus on decarbonization and operational efficiency in port systems (Alamoush et al., 2021; Ferrari et al., 2023). Two reinforcing transitions are reshaping port logistics. The first is digitalization, where Internet of Things (IoT) architectures link cranes, vehicles, containers, buildings, and utility infrastructure through pervasive sensing and wireless communications. These data streams enable situational awareness, dynamic scheduling, and automated decision support that raise throughput reliability and reduce delay (Belmoukari et al., 2023; Inkinen et al., 2019; Pham, 2023).

The second is the energy transition inside the port perimeter. Onshore power supply allows berthed vessels to shut down auxiliary engines, cutting local emissions when grid electricity is low carbon. Complementary measures include electrified handling equipment, port microgrids with storage, and early green hydrogen initiatives for heavy duty applications. Evidence shows substantial reductions in pollutants at berth and growing standardization and feasibility across ports, although business models and grid reinforcements remain decisive (Abu Bakar et al., 2023a; Elkafas & Seddiek, 2024; Innes & Monios, 2018). Case-led analyses also document pathways for port authorities to catalyse renewable deployment and coordinate actors, as illustrated by Rotterdam's energy programmes and related governance arrangements (Hentschel et al., 2018). Integrating these transitions multiplies benefits. Digital twins and data analytics can align berth scheduling, equipment dispatching, and charging strategies with variable renewable generation and storage constraints, improving both operational and energy efficiency. Research on port microgrids and operations planning demonstrates how joint optimization smooths peaks, lowers energy costs, and supports reliability in high-density terminals, including contexts inspired by Singapore's container operations (Iris & Lam, 2021; Yap et al., 2023).

This chapter synthesizes these strands to present an integration framework that links the physical layer of energy and logistics assets, the digital layer of sensing and

DOI: 10.1201/9781003671039-3

analytics, and the decision layer of governance and market rules. It then examines barriers common to implementation, including compatibility across vendors, investment risks, data and cybersecurity requirements, and multi-stakeholder coordination, before outlining policy and managerial levers to deliver scalable outcomes in diverse port settings.

3.2 LITERATURE REVIEW

3.2.1 IoT in Port Logistics and Smart Port Development

The notion of the smart port has matured from an aspiration to a broad research agenda that links operational excellence with sustainability outcomes. Recent systematic reviews describe smart ports as complex socio-technical systems in which digital technologies orchestrate assets, actors, and information to improve efficiency, safety, and environmental performance. These reviews also trace the field's evolution from mechanized cargo handling towards fifth-generation ports that embed connectivity, sensing, and analytics across terminal processes and community interfaces (Belmoukari et al., 2023). Within this trajectory, bibliometric syntheses converge on the centrality of the IoT as the connective layer that binds equipment, facilities, and information platforms, while also mapping dominant themes such as data governance, interoperability, and human capital (Pham, 2023).

Empirical and review evidence shows that port IoT architectures now span quay cranes, yard equipment, gate systems, warehouses, and utilities, streaming high frequency telemetry that supports real-time situational awareness and coordinated operations. These data flows feed decision support for berth alignment, yard planning, truck appointments, and inventory control, with reported gains in reliability and service speed (Pham, 2023). A growing body of work focuses on condition monitoring and predictive maintenance, where vibration and electrical signatures from critical subsystems are mined to detect anomalies and trigger maintenance before failure; studies on quay crane motors exemplify this shift from periodic to condition-based interventions that reduce unplanned downtime (Mukherjee et al., 2024). Parallel advances in digital information sharing emphasize how Port Community Systems aggregate IoT-enabled data across organizational boundaries; structured reviews underline the need for shared data models and governance rules to avoid silos and to enable synchronized transactions among lines, terminals, forwarders, and authorities (Moros-Daza et al., 2020).

Environmental monitoring is another prominent strand. Research documents the influence of port activity on local air quality and public health, strengthening the rationale for continuous sensing of pollutants and micro-meteorology in and around port precincts and for using these data in compliance and mitigation planning (Ducruet et al., 2024). Across these domains, an important insight from recent syntheses is that IoT delivers most value when paired with analytics and platformization. Studies highlight how fused telemetry location, utilization, vibration, energy, and environmental streams support predictive and prescriptive models for scheduling, risk management, and resource allocation along the maritime supply chain (Surucu-Balci et al., 2024).

The literature also maps constraints that shape scaling. Interoperability remains a first-order concern because legacy systems and vendor-specific protocols fragment data and impede end-to-end visibility; reviews of Port Community Systems and digital transformation call for standards alignment, open interfaces, and clear data ownership rules (Moros-Daza et al., 2020; Surucu-Balci et al., 2024). Cybersecurity emerges as a parallel design constraint. Sector-specific analyses for European ports and systematic reviews of maritime cybersecurity point to expanding attack surfaces and capability gaps, reinforcing the need for authentication, segmentation, continuous monitoring, and incident response that are embedded in project design rather than appended after deployment (Bocayuva, 2021; Bolbot et al., 2022). Finally, work on distributed ledgers and document digitization positions blockchain as a potential trust mechanism for multi-party data and document exchange, while cautioning that benefits depend on stakeholder alignment and regulatory clarity (Liu et al., 2023).

Taken together, the literature positions IoT as both sensing infrastructure and coordination medium. Realizing its benefits at scale depends on interoperable data ecosystems, analytics capability, and security and governance arrangements that translate continuous data into coordinated action within terminals and across the wider port community.

3.2.2 Renewable Energy Deployment in Ports and Green Port Strategies

Parallel to digitalization, ports are pursuing "green port" strategies that aim to reduce environmental impacts through cleaner energy sources and improved efficiency. Tawfik and colleagues present a case study of a Mediterranean port where solar panels installed on port buildings generated enough electricity to meet the buildings' consumption, while simulated wind arrays exhibited high power coefficients for future implementation (Tawfik et al., 2023). Their optimization analysis showed that combining solar and wind systems can support energy needs and improve the port's environmental footprint. More broadly, ports in sunny regions leverage rooftop photovoltaic installations and solar carports, whereas ports with favourable wind regimes deploy onshore or near-shore turbines. The integration of renewable sources diversifies energy supply and reduces reliance on fossil fuels; however, it requires careful management due to the variability of solar and wind output.

The literature increasingly views ports not merely as energy consumers but as integrated energy systems. Yu and colleagues explain that integrated port energy systems (IPES) are characterized by multi-energy flow coupling and "physics, information and value" interaction, where electricity, hydrogen, thermal energy, and logistics activities are tightly interlinked (Tawfik et al., 2023). Because the coupling relationships are complex and external factors such as weather and market conditions are uncertain, traditional engineering models cannot adequately capture system dynamics (Yu et al., 2023). Digital twin technology has emerged as a promising solution: by creating high-fidelity virtual models of port energy systems, it can simulate scenarios, provide real-time feedback, and support low-carbon operation (Sifakis et al., 2021; Yu et al., 2023). The digital twin approach complements renewable integration by modelling how energy from solar panels, wind turbines, and hydrogen generators interacts with port logistics and by identifying optimal strategies for energy storage and dispatch. As ports electrify equipment and adopt zero-carbon fuels, integrated energy systems will increasingly adopt microgrids that combine main grid connections with local distribution and storage, enabling ports to function as independent energy regions (Yu et al., 2023).

Cold ironing also termed shore power represents another critical sustainability measure. By providing vessels with electricity from the grid or on-site renewables while at berth, ports can eliminate emissions from auxiliary engines. Studies by Chang, Wang, and other researchers demonstrate that shore power can reduce carbon dioxide emissions by more than 50% and particulate matter by about 40% (Tawfik et al., 2023). Cold ironing also reduces noise and vibration, benefiting nearby communities. Nevertheless, adoption remains uneven. A survey of Vietnamese port managers by Le finds that economic incentives and regulatory support positively influence cold ironing uptake, whereas high capital costs and lack of standardization are barriers (Le, 2024). In developed regions, ports such as Los Angeles and Rotterdam have integrated shore power for container ships and cruise vessels, drawing electricity from renewable sources to maximize environmental benefits.

Hydrogen is emerging as a promising energy carrier for ports. A report by the U.S. National Renewable Energy Laboratory identifies demonstration projects at the Port of Los Angeles, Port of Valencia, and planned trials in Hamburg where hydrogen fuel cells power yard tractors and port equipment (Barberi et al., 2022; Le, 2024). These initiatives show that hydrogen can act as a range extender for battery-dominant vehicles, providing zero-emission operation for heavy-duty equipment. Such projects are part of broader efforts to develop hydrogen bunkering and refuelling infrastructure for vessels. Ports are also exploring renewable ammonia and methanol as future fuels, yet these remain in early stages.

In addition to energy supply, ports are electrifying their internal vehicle fleets forklifts, terminal tractors, and cranes, and installing charging infrastructure. Electrification reduces fuel consumption and local emissions; however, it increases electricity demand and underscores the need for smart microgrids that manage load peaks and coordinate charging. Digital twin technology offers a solution. Yu and co-authors argue that digital twins provide high-fidelity virtual models of IPES, enabling real-time data interaction, and optimization (Yu et al., 2023). Digital twins support the "physics-information-value" coupling of port operations by mapping physical assets to digital models and facilitating predictive control. They also enable ports to balance energy supply and demand across multi-energy networks, including electricity, hydrogen, and thermal systems (Yu et al., 2023). These capabilities make digital twins a key component of next-generation green ports.

3.2.3 Towards Integrated Frameworks in Developing Economy Ports

Blockchain and artificial intelligence (AI) are also discussed as smart and green port enabling technology in the literature. Tijan et al. (2019) gave an extensive overview of how blockchain can be used in shipping industry, noting its high likelihood to increase the level of transparency, traceability, and efficiency on the part of supply chain management. Blockchain is used to refer to a shared ledger or record that uses data transaction in a decentralized and tamper immune system, where the record of the ledger is shared between nodes. Blockchain has the potential to minimize paper effort by making records of cargo transportation, contracts, and compliance to accord permanent records (Farah et al., 2024). The survey highlights that integrating blockchain into maritime supply chains could resolve long-standing challenges of document traceability and provenance verification. Farah et al. (2024) propose roadmaps for adoption that involve assessing port network infrastructure and aligning blockchain platforms with existing IoT systems. They also identify cybersecurity risks, including double-spending attacks, smart contract bugs, and vulnerabilities in consensus mechanisms. To mitigate these issues, researchers suggest consortium or private blockchains with robust authentication protocols and data governance models, while recognizing that decentralized ledger technology must be reconciled with data protection laws (Farah et al., 2024).

There is an emerging use of AI in maritime logistics and in environmental monitoring activities. Prousaloglou et al. (2025) reveal that machine learning, natural language processing, etc., are AI applications capable of manipulating high data, optimizing cargo handling, forecasting vessel routing, and fuel consumption. They insist that AI can lighten day-to-day port operations, especially when it comes to loading and unloading activities, as it will automate decision-making and allow stowage plans to be made more accurately. It also points out the use of generative AI tools in communication and decision-making in real time: natural language processing is also useful in coordination among stakeholders and quick access to the market to manage risks. AI also supports environmental sustainability by enhancing emissions monitoring, ecological impact assessment, and compliance with regulations (Prousaloglou et al., 2025). Integrating AI with IoT sensors and digital twin models yields powerful predictive capabilities for fuel consumption, route optimization, and maintenance planning. As digitalization advances, AI is poised to become a cornerstone of smart port strategies, turning data into actionable knowledge and enabling autonomous or semi-autonomous operations.

Farah et al. (2024) underline that the primary contributions of the blockchain to maritime logistics highlighted by them are increased transparency and traceability, potential frameworks of integrating blockchain into the current port activities, evaluating the level of network infrastructure readiness, and reviewing certain cyber threats and challenges. Some use cases presented in the survey report on Maersk and IBM TradeLens platform indicating how real-time access to shipping data and documents via a decentralized ledger can be used to automate processes and eliminate fraud. The study also highlighted about similar uses of blockchain in other sectors which demonstrate its efficiency in making sure that the records in the supply chains are fair (Farah et al., 2024). Such conclusions are practically oriented in that they provide guidance to ports on the possible deployment of blockchain, where they should consider measuring the network capacity, identifying the right consensus mechanism, and ensuring gradual adoption into the existing digital ecosystems. Table 3.1 presents a thematic synthesis of scholarly contributions at the intersection of IoT applications, smart port logistics, and renewable energy integration. It categorizes the existing literature into eight key research themes, highlighting focus areas,

TABLE 3.1
Key research themes in smart port logistics and renewable integration

Theme	Key Focus Areas	Sources	Identified Gaps
Digitalization in Port Operations	RFID, AIS, yard automation, digital twins, predictive maintenance	Heikkilä et al. (2022); Jin & Guo (2025); Jović et al. (2022); Klar et al. (2023); Zerbino et al. (2019)	Most studies based on high-income countries; limited applicability in informal or fragmented port environments
Renewable Energy Integration	Solar panels, wind farms, tidal energy, cold ironing, smart lighting	Cavalli et al. (2021); Philipp et al. (2021)	Technical feasibility established, but economic and institutional feasibility underexplored
IoT in Energy Monitoring	Energy demand prediction, smart metres, cloud-based dashboards	Dzemydienė et al. (2023); Yigit et al. (2023)	Real-time feedback mechanisms poorly integrated with traditional port infrastructure
Microgrid and Shore Power Management	Electrification of ships at berth, port-side grid stability, energy storage	Dinh et al. (2024); González-Cancelas et al. (2024)	Most models not optimized for variable renewable energy inputs; interoperability with local utilities unclear
Sustainability-Oriented Port Logistics	Low-carbon logistics chains, environmental KPIs, zero-emission targets	Cavalli et al. (2021); Heikkilä et al. (2022)	Limited port-specific sustainability benchmarks; often excludes freight hinterland integration
Implementation Challenges in Emerging Economies	Infrastructure ageing, digital skill gaps, low R&D investment, public–private misalignment	Jin Jiahui & Guo (2025); Zaychenko et al. (2021)	Lack of bottom-up planning and stakeholder inclusion frameworks
Data Governance and Cybersecurity	Digital trust, port data platforms, threat detection, information sharing	Alahmadi et al. (2021); Ouyang et al. (2023)	Absence of unified data standards and regulatory compliance frameworks
Policy and Multi-Level Coordination	Green port policies, smart maritime corridors, national innovation plans	Alam et al. (2016); Lacalle et al. (2020); Molavi et al. (2020)	Disconnect between maritime innovation agendas and climate targets at regional level

representative peer-reviewed studies, and identified research gaps. The table reveals a strong concentration of empirical work in high-income port economies, limited integration of IoT with renewable energy systems in real-time operations, and persistent challenges in data governance and policy alignment, particularly in emerging economies.

3.3 IOT IN PORT LOGISTICS: CONCEPTS AND TECHNOLOGIES

3.3.1 IoT Architecture and Components

IoT architectures in ports are typically organized into three layers: the perception layer, the network layer, and the application layer (Wang & Hsu, 2025). The perception layer comprises sensors, actuators, and radio-frequency identification (RFID) tags attached to physical assets such as containers, cranes, yard vehicles, and infrastructure. Sensors capture data on location, movement, temperature, humidity, vibration, energy use, and environmental conditions (Wang & Hsu, 2025). Intelligent devices integrated into cargo holds, warehouses, and vehicles collect continuous streams of operational data that serve as the foundation for subsequent analysis (Wang & Hsu, 2025). The network layer provides connectivity through technologies such as Wi-Fi, 5G, long-range wide-area networks (LoRaWAN), and satellite links. This layer aggregates sensor data and transmits it to servers or cloud platforms. In modern implementations, edge computing units installed at terminals pre-process data to reduce latency and network load before forwarding critical information to the cloud. The application layer includes data analytics platforms, Port Community Systems, and decision support tools that process the collected data, generate insights, and orchestrate actions. By integrating descriptive statistics, clustering, forecasting, and association rule mining, these analytics platforms extract patterns and predict future states (Wang & Hsu, 2025). Automated alerting systems use these insights to inform operators of equipment malfunctions or scheduling conflicts. In practice, the three layers are interconnected through middleware and API gateways, allowing interoperability among devices from multiple vendors.

Figure 3.1 illustrates the broader landscape of IoT, highlighting its evolution, definition, logistics applications, and future prospects. This conceptual overview complements the three-layer IoT architecture by emphasizing the transformative role of IoT in logistics through real-time tracking, intelligent warehousing, and predictive management strategies (Wang & Hsu, 2025).

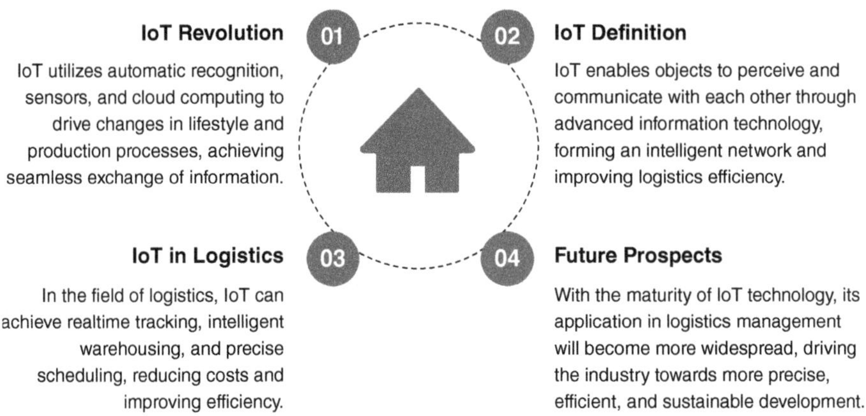

FIGURE 3.1 Overview of the Internet of Things (IoT).

Source: Adopted from (Wang & Hsu, 2025)

3.3.2 Operational Applications

IoT technologies enable a wide range of operational applications in ports. Real-time tracking of vessels and containers allows port authorities and logistics providers to monitor ship arrivals, berthing schedules, cargo discharging, and gate operations. Wang and Hsu (2025) emphasize that continuous monitoring of cargo status and environmental parameters through IoT devices enhances logistics efficiency and safety. Smart sensors installed in storage areas monitor temperature and humidity to ensure appropriate conditions for sensitive goods (Aslam et al., 2025; Tijan et al., 2019). Integrated monitoring also supports automated identification of hazardous materials and regulatory compliance.

Predictive maintenance is another key application. Sensors embedded in cranes and gantry systems measure vibration, torque, and operating cycles to detect anomalies and predict component failures. By analysing this data, maintenance teams can schedule repairs before equipment breakdowns occur, reducing downtime and improving asset utilization. IoT-enabled predictive maintenance not only increases operational reliability but also extends the service life of expensive machinery (Muñuzuri et al., 2020).

IoT platforms also optimize yard management and traffic flow. Automated terminal operating systems use sensor data to assign trucks, allocate stacking slots, and sequence container movements. Real-time analytics identify congestion hotspots and adjust gate hours or truck appointments accordingly. In Rotterdam, for example, IoT-based traffic management systems coordinate vessel arrivals with hinterland transportation to minimize wait times and reduce emissions (Li et al., 2023). IoT cameras and computer vision algorithms enable automated container inspection, damage detection, and security monitoring, increasing throughput and security.

3.3.3 Data Analytics and AI

The explosion of data from IoT devices necessitates advanced analytics. Gerakoudi et al. (2024) employ machine learning to review the maritime IoT literature and observe that IoT technology transforms the shipping industry into an interconnected digital environment capable of processing information automatically. Integrated data flows facilitate global, real-time connectivity among market participants and support evidence-based decision-making. AI algorithms analyse patterns in vessel arrivals, cargo flows, and equipment utilization, enabling dynamic berth allocation, container stacking optimization, and demand forecasting. Data analytics also improve safety by detecting anomalies in vessel movements and enabling predictive collision avoidance.

Moreover, AI assists in reducing energy consumption. By analysing historical energy usage and operational patterns, AI can optimize lighting, heating, ventilation, and air-conditioning systems in port buildings. It can also coordinate charging of electric vehicles and operation of battery storage to flatten demand peaks. Prousaloglou et al. highlight that AI tools can process vast datasets to improve cargo management, route forecasts, and environmental monitoring (Prousaloglou et al., 2025). Natural language processing facilitates communication among stakeholders and enhances situational awareness during complex operations (Li et al., 2023; Prousaloglou et al., 2025). In addition, recent research stresses the importance of data analysis methods in maritime logistics management; descriptive statistics, clustering, and forecasting techniques help identify bottlenecks, optimize routing, and improve scheduling (Li et al., 2023; Muñuzuri et al., 2020). Visualization tools provide intuitive overviews of complex supply chains, enabling operators to detect anomalies and make timely interventions (Zaychenko et al., 2021). Combining AI with IoT-generated data therefore transforms ports into data-driven environments where predictive and prescriptive analytics underpin decisions on vessel arrival times, berth allocation, and inventory management.

3.3.4 Interoperability and Skills Challenges

Despite these advantages, IoT implementation faces several challenges. Integrating heterogeneous devices from different vendors requires common standards for communication and data formats. Without interoperability, ports risk creating isolated systems that cannot share data effectively. Gerakoudi et al. (2024) and Tijan et al. (2019) argue that IoT adoption is impeded by insufficient standardization and limited infrastructure capable of supporting high-bandwidth wireless networks. Data quality and analytics capacity are also concerns; transforming raw sensor data into actionable information demands skilled data scientists and engineers. The maritime sector often lacks personnel trained in quantitative analysis and IoT technologies (Li et al., 2023; Tijan et al., 2019). This skills gap limits the ability of shipping companies and port authorities to fully exploit IoT-generated data. Finally, cybersecurity remains a critical issue, as increased connectivity exposes ports to potential cyberattacks. Strategies to mitigate cyber risks include robust encryption, network segmentation, regular vulnerability assessments, and staff training.

3.4 RENEWABLE ENERGY INTEGRATION IN PORTS

3.4.1 Solar, Wind, and Emerging Renewable Sources

Solar and wind power are the most mature renewable technologies adopted by ports. Ports often possess large roof surfaces on warehouses and terminals that can host photovoltaic panels. In Egyptian ports, Tawfik et al. demonstrate that installed solar panels can generate sufficient energy to power port buildings, while simulated wind arrays exhibit favourable power coefficients for future implementation (Tawfik et al., 2023). Many ports in Europe and Asia have installed rooftop solar farms or ground-mounted arrays to offset electricity consumption. For instance, Jurong Port in Singapore operates one of the world's largest rooftop solar installations on its warehouses. Coastal ports also take advantage of strong wind resources by installing turbines on site. The Port of Rotterdam and several UK ports have onshore turbines that supply electricity to port facilities and feed surplus power to the grid. Combining solar and wind generation diversifies the energy mix and enhances reliability because solar output tends to peak during daytime while wind speeds can be higher at night.

3.4.2 Cold Ironing (Shore Power) and Electrification of Ships

Beyond solar and wind, ports are exploring tidal and wave energy technologies that harness ocean currents and waves. Although these projects are in pilot stages, they offer predictable generation patterns that complement variable solar and wind resources. Some ports have installed small tidal turbines in harbour channels to demonstrate the feasibility of marine energy. Green hydrogen production through electrolysis is another promising avenue (Barberi et al., 2022). Ports can use excess renewable electricity to produce hydrogen, which serves as an energy storage medium and fuel for heavy-duty vehicles and equipment. The U.S. National Renewable Energy Laboratory reports ongoing demonstration projects at the Port of Los Angeles and the Port of Valencia where hydrogen fuel cells power yard tractors and extend the range of battery-dominant equipment. Such projects aim to establish hydrogen as a viable alternative to diesel in port operations. Ports are also examining the construction of hydrogen bunkering facilities to supply future hydrogen-powered vessels (Abu Bakar et al., 2023b; Le, 2024).

3.4.3 Smart Microgrids and Energy Storage in Ports

Cold ironing provides shore electricity to berthed vessels, enabling them to shut down auxiliary engines and thereby reducing emissions and noise. Numerical studies suggest that reducing ship speed near ports and supplying onshore power can cut carbon dioxide emissions by more than 50% and particulate matter by nearly 40% (Tawfik et al., 2023). Le's survey of Vietnamese ports indicates that economic incentives and regulatory mandates increase cold-ironing adoption, whereas high installation costs and lack of standardization impede it (Tawfik et al., 2023). In Europe and North America, many ports have installed shore power facilities that rely on renewable energy or purchase green electricity from the grid. The Port of Los Angeles reports that cold ironing has reduced local particulate pollution by more than 50% compared with levels in the mid-2000s.

Electrification extends beyond vessels to port equipment. Battery-electric forklifts, yard tractors, straddle carriers, and gantry cranes are becoming more common as their performance improves and costs decline. Some terminals have adopted hydrogen fuel cell versions of these vehicles. Electrification reduces on-site emissions but requires robust charging infrastructure and careful load management to avoid grid congestion. Digital twin technology can help coordinate charging schedules and integrate renewable generation and battery storage. Yu and colleagues argue that digital twins can model the dynamic interplay between energy supply and demand, optimize energy flows, and support decision-making for IPES (Yu et al., 2023). As ports electrify both ships and equipment, the ability to orchestrate energy resources becomes critical.

3.4.4 Institutional and Policy Barriers to Renewable Integration

Renewable integration necessitates sophisticated energy management. Microgrids allow ports to operate semi-independently from the main grid, combining on-site generation, battery storage, and controllable loads. IoT sensors gather data on generation, consumption, and storage levels, feeding it into energy management systems that optimize dispatch. Digital twins of port energy systems simulate various scenarios and control strategies, improving resilience and enabling predictive maintenance. Such systems can prioritize renewable consumption, shift loads to off-peak periods, and coordinate with grid operators for demand-response. Beyond serving as backup power supplies, microgrids are becoming the backbone of IPES. Recent research notes that electrification and zero-carbon fuels such as hydrogen and ammonia promote the integration of multiple energy networks electricity, gas, and thermal within ports (Yu et al., 2023). In this vision, ports act as independent energy regions connected to the main grid via "main power grid + distribution grid + microgrid" structures (Yu et al., 2023). By participating in wholesale electricity markets, ports can provide ancillary services and generate additional revenue while ensuring reliable shore power for vessels. Integrated energy management also supports the electrification of transport modes connecting to the port rail, trucks, and barges and aligns with broader smart city initiatives.

Despite these opportunities, renewable integration also introduces new risks. Cybersecurity concerns arise as increased digital connectivity expands the attack surface of port energy systems. Reliability issues emerge from dependence on variable renewable resources, which may create vulnerabilities in balancing supply and demand. Safety risks are also significant, particularly in handling hydrogen and ammonia fuels, managing high-capacity batteries, and ensuring grid stability. Addressing these barriers requires robust cybersecurity frameworks, redundant backup systems, and stringent safety protocols to complement technical integration and policy measures.

3.5 INTEGRATION FRAMEWORK AND GOVERNANCE MODELS

Effective integration of IoT and renewable energy requires a holistic framework that aligns physical infrastructure, digital systems, and governance mechanisms. Building on the literature, a three-layer framework can guide implementation. Figure 3.2 represents the integration of physical infrastructure, digital platforms, and governance mechanisms, capturing the flow from IoT-enabled energy systems to coordinated decision-making.

3.5.1 Physical Layer

The physical layer encompasses IoT devices, renewable energy assets, and infrastructure. Sensors and actuators are attached to cranes, vehicles, containers, buildings, and energy systems (solar panels, wind turbines, hydrogen generators). These devices collect data on operational status, energy production, weather conditions, and environmental parameters. The physical layer also includes power distribution systems, microgrids, and energy storage units. Ensuring reliability at this level requires resilient design, redundant power supplies, and adequate protection against harsh marine environments. Equipment must be compatible with IoT connectivity and able to interface with digital systems.

3.5.2 Digital Layer

The digital layer processes and analyses data are generated by the physical layer. It comprises communication networks, cloud or edge computing platforms, data warehouses, digital twins, and analytics tools. Aslam et al. (2025), Yang et al. (2018),

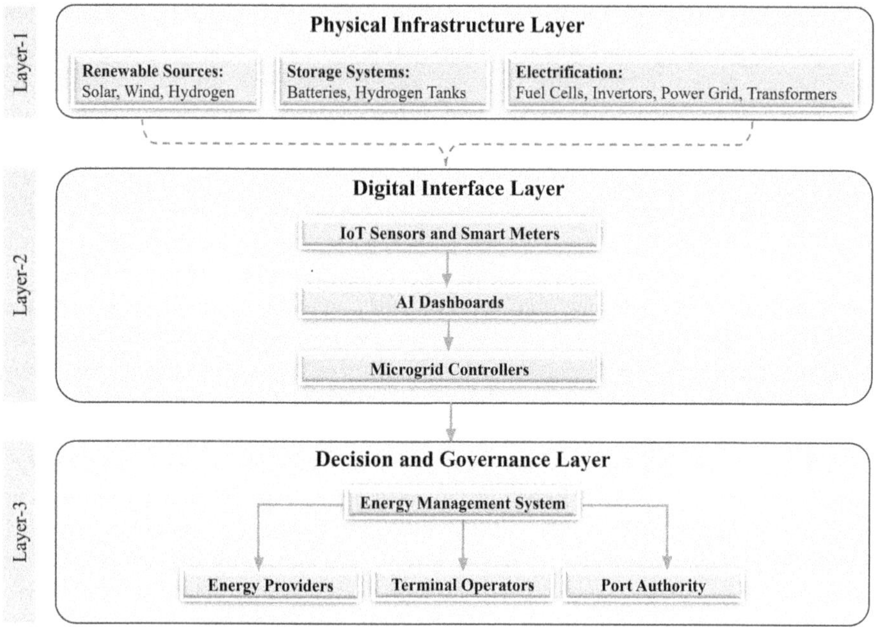

FIGURE 3.2 Integrated governance and technology framework for smart port energy systems.

and Yu et al. (2023) describe digital twins as virtual replicas of IPES that enable high-fidelity modelling, real-time data interaction, and closed-loop optimization. Unlike traditional simulations, digital twins are pre-trained models that are continuously refined through real-time data exchange and feedback (Yu et al., 2023). This dynamic update capability allows the twin to approximate the physical system with high fidelity and support intelligent decision-making. Digital twins integrate information flows across multiple energy sources and port operations, allowing operators to simulate scenarios, forecast demand, and adjust operations accordingly (Yu et al., 2023). They also facilitate coupling between energy networks and logistics processes by modelling interactions across the physical, informational, and economic dimensions. Blockchain networks may be incorporated to ensure secure and immutable data sharing among stakeholders (Farah et al., 2024), while AI algorithms identify patterns, detect anomalies, and optimize scheduling (Prousaloglou et al., 2025). Data visualization dashboards provide decision-makers with intuitive insights. As digital platforms evolve, ports can adopt modular architectures in which microservices manage specific functions, enabling scalability and flexibility. Interoperability at the digital layer hinges on open interfaces and standard data formats that link devices, energy systems, and software platforms.

3.5.3 Decision-Making Layer

The decision-making layer encompasses governance structures, policies, operational procedures, and human expertise. This layer interprets insights from the digital layer and orchestrates actions in the physical layer. It includes strategic planning (e.g., capacity expansion, equipment procurement), operational control (e.g., adjusting berth schedules, dispatching maintenance teams), and regulatory compliance. Effective decision-making requires coordination among port authorities, terminal operators, shipping lines, energy providers, and government agencies. Policies should align incentives for digitalization and renewable investment, establish common standards, and enforce cybersecurity and privacy safeguards. The decision-making layer also involves training personnel in data analytics, cyber hygiene, and energy management.

3.6 GOVERNANCE AND POLICY DIMENSIONS

Implementing IoT and renewable energy at ports implicates a wide array of governance and policy issues. Regulatory frameworks determine how data is collected, shared, and protected, and they shape market conditions for renewable investment and infrastructure financing. The complexity of IPES underscores the need for coordinated policy and research frameworks. Yu et al. point out that existing theoretical systems for cross-domain and multi-network coupling are lacking, making it difficult to conduct systematic and in-depth research on port energy systems (Tijan et al., 2019; Yu et al., 2023). The absence of unified analytical models can hinder evidence-based policymaking and standard development. Policymakers should therefore support interdisciplinary research programmes that develop modelling tools, standards, and regulatory guidelines for integrated energy and logistics systems. Such initiatives will provide the scientific foundation for regulations governing microgrids, hydrogen infrastructure, and digital twins, and will ensure that innovations are compatible with climate targets and energy security goals.

3.6.1 Data Governance and Privacy

IoT deployment produces vast volumes of operational and personal data. Ensuring that this data is managed responsibly is essential for public trust and legal compliance. The European Union's General Data Protection Regulation (GDPR) requires organizations to protect personal data and provide rights such as data erasure. However, the immutable nature of blockchain can conflict with these principles. Ben Farah et al. explain that blockchain replicates data across multiple nodes, making it difficult to fulfil "right to be forgotten" requests (Yu et al., 2023). Ports adopting blockchain must therefore design systems that balance transparency and privacy, for example, by storing sensitive data off chain or using privacy-preserving cryptographic techniques. Ports should implement data classification schemes, anonymization where appropriate, and clear data ownership agreements among stakeholders.

3.6.2 Standards and Interoperability

Lack of common standards hinders the integration of IoT devices and renewable systems across different terminals and jurisdictions. International bodies such as the International Organization for Standardization (ISO) and the International Electrotechnical Commission (IEC) are developing standards for IoT architectures, data formats, cybersecurity, and energy management. Ports should adopt these standards to ensure interoperability and ease of integration. Government agencies can encourage standard adoption by tying compliance to funding or permitting conditions. Collaborative platforms like Port Community Systems and digital corridors can facilitate data exchange among ports, shipping lines, and hinterland transportation modes, enhancing end-to-end visibility.

3.6.3 Incentives for Renewable Investment

High capital costs remain a primary barrier to renewable energy projects such as cold ironing and large-scale solar installations. Le (2024) found that economic incentives and regulatory support positively influence adoption. Carbon pricing and emissions trading schemes also strengthen the business case for renewable investment. At the same time, regulations can mandate emission reductions and require ships to use shore power or alternative fuels when available. Aligning incentives and mandates encourages ports to invest in green energy infrastructure while providing certainty to shipping companies.

3.6.4 Stakeholder Collaboration

Ports operate at the intersection of multiple stakeholders: terminal operators, shipping lines, logistics providers, energy companies, municipalities, and communities. Integrating IoT and renewable energy requires coordinated actions and clear delineation of roles. Stakeholder platforms can facilitate dialogue on technology standards, data sharing, environmental targets, and investment priorities. Partnerships between public and private actors can mobilize funding and expertise for large projects, such as hydrogen hubs or digital twin development. Port governance must also consider environmental justice, ensuring that local communities benefit from reduced pollution and access to cleaner energy.

3.7 POLICY RECOMMENDATIONS AND STRATEGIC ROADMAP

Transforming a conventional port into a smart, renewable-integrated port requires not only technology deployment but also supportive policies and a clear implementation strategy. Based on the analysis, we outline several policy recommendations and a phased roadmap to guide stakeholders through this transition. The recommendations target port authorities, government policymakers at national and local levels, as well as industry associations.

3.7.1 Policy Recommendations

3.7.1.1 Develop and Enforce Standards and Interoperability Protocols

To ensure different systems can work together, standards are essential. Governments and port authorities should mandate or strongly encourage the use of open standards for data and IoT in ports. For example, adopting a common data model for Port Community Systems (as per IMO's FAL Convention requirements) and ensuring any new digital equipment is compliant with industry communication standards (like IMO/IEC standards for shore power connectors, and ISO standards for container identification and tracking). A national or regional guideline for "Smart Port Standards" could be formulated, possibly by a coalition of port authorities under the guidance of a transport ministry or standards body. This would cover everything from messaging formats (for vessel and cargo data exchange) to technical standards for automation interfaces. As a policy, this reduces vendor lock-in and makes it easier for smaller ports to piggy-back on proven solutions from larger ports. Standardization was identified by Inkinen et al. (2021) as a key driver for port digitalization success, and similarly, interoperability is crucial for scaling renewable integration across ports. Additionally, governments can support the development of reference architectures or pilot implementations that ports can emulate, reducing uncertainty about what technologies to choose.

3.7.1.2 Capacity Building and Training Programmes

Investing in human capital is as important as investing in technology. Port workers, managers, and regulators need new skills to handle IoT systems, data analytics, and renewable energy operations. We recommend establishing training programmes and knowledge exchange initiatives possibly via port training institutes or partnerships with universities/polytechnics. These programmes should cover digital literacy for port operations, cybersecurity best practices, and technical training for maintaining electric and automated equipment. For instance, a "Smart Port Academy" could be created at major ports to continuously train staff from ports around the country (or region), supported by government funding. Multilateral bodies like UNCTAD and the World Bank are already focusing on this; UNCTAD's TrainForTrade programme, for example, includes a module on port management in the digital era. National governments can allocate funds for upskilling port labour so that the workforce transition is smooth and inclusive, addressing concerns of job displacement by reorienting workers to new roles (e.g., an electrical technician course for mechanics transitioning to maintain electric vehicles). Without such proactive measures, technology adoption may face resistance or underperformance due to skill gaps.

3.7.1.3 Financial Incentives and Funding Mechanisms

To overcome the high upfront costs, a mix of incentives and innovative financing is needed. Policymakers can introduce tax credits or accelerated depreciation allowances for investments in green port equipment and IoT infrastructure. For example, a port installing a solar plant or cold ironing facility could get a tax rebate or a subsidy covering a percentage of the cost. Grants or low-interest "green loans" can be provided for pilot projects demonstrating smart and sustainable port solutions, especially at publicly owned ports. Creating green port investment funds, possibly as public–private funds – can pool capital for multiple ports to borrow from for such upgrades. International climate finance should be tapped: many port projects qualify for climate mitigation funding (since they reduce emissions), and development banks might co-finance these. At the same time, adjusting the tariff structures at ports can encourage private operators to invest: port authorities could offer reduced concession fees or extended lease terms for terminal operators that meet certain technology and emissions benchmarks (for instance, if a terminal operator achieves a set reduction in carbon emissions by using electrified equipment, they get a lease extension or a fee reduction as a reward). This creates a direct business incentive to modernize. On the demand side, ports can incentivize users, for example, offering discounted port dues for ships that are IoT-integrated with the port's scheduling system (improving efficiency) or that use shore power or low-sulphur fuel (improving emissions), as some ports already do via Green Port programmes (Tawfik et al., 2023). Essentially, internalizing environmental and efficiency benefits into the port pricing signals helps align private decisions with the smart/green objectives.

3.7.1.4 Strengthen Institutional Coordination and Governance Frameworks

Given that integration requires multiple stakeholders to cooperate, formal governance mechanisms should be established or enhanced. We recommend forming Smart Port Coordination Committees at each major port, chaired by the port authority, including representatives from terminal operators, shipping lines, trucking companies, utility providers, and local government. These committees would meet regularly to share data, set targets (like "20% energy from renewables by 2025" or "average truck turnaround <30 minutes"), and monitor progress. They can also jointly troubleshoot issues, for instance, if a new truck appointment system isn't being used, the committee can identify why and how to improve compliance. On a national level, a Port Digitalization and Sustainability Task Force can align policies across ports and ensure that lessons learned are disseminated. This could be under the Ministry of Ports/Transport. For countries with many ports, a phased rollout might be coordinated by a pilot in one flagship port, then replicated in others with central support. Moreover, integrate port strategies with broader national strategies (e.g., if a country has a renewable energy policy or an EV policy, ensure ports are explicitly included). The case of India's Maritime India Vision 2030, which calls for transitioning Indian ports to "world-class Smart Ports" by 2025, is a good example, implementing it will require continuous oversight and coordination (Chen et al., 2019; Inutsuka et al., 2024).

3.7.1.5 Regulatory Measures for Emissions and Data Sharing

Some aspects may require the stick in addition to the carrot. For environmental goals, regulations can ensure compliance: for example, mandating that by a certain year, all handling equipment procured in ports must be zero-emission (electric or hydrogen) where feasible, or setting emissions caps for port operators. California's Air Resources Board has done this, requiring cargo handling equipment in ports to gradually transition to zero emissions by 2030. Such regulations drive innovation and investment by providing certainty. Similarly, to ensure data-sharing, governments might need to mandate usage of national single-window or Port Community Systems for logistics data, as has been done with customs single windows. If each stakeholder is required to submit and exchange data through a common platform (with appropriate confidentiality safeguards), the port as a whole becomes more transparent and efficient. A light-touch regulatory approach could first encourage and then require integration of systems, for example, first provide incentives for ships to share real-time arrival data via the port's system; later, require advance electronic data submission as a condition of port entry (some ports and IMO's e-nav guidelines already do this). A balanced approach combining voluntary compliance and eventually compulsory standards tends to work, as seen in the evolution of electronic customs documentation globally.

3.7.1.6 Cybersecurity and Resilience Planning

Policymakers and port authorities should ensure robust cybersecurity frameworks are in place alongside any digital rollout. This means adopting international cybersecurity standards in port operations, conducting regular audits and penetration testing, and sharing threat intelligence through industry groups (e.g., IMO guidelines on maritime cyber-risk management). Policy could require ports above a certain size to have a cyber-risk management plan and incident response protocol. Additionally, incorporating redundancy and resilience into plans, for instance, maintaining backup power for critical systems (some ports keep diesel generators for essential operations in case the grid or microgrid fails) and fallback manual procedures if digital systems go down, to avoid total paralysis in an outage. Drills and training in these contingency procedures should be mandated to ensure readiness.

3.7.1.7 Community and Environmental Policies

Finally, smart and green port policies should be integrated with local community development. Ports should engage with local stakeholders when planning renewable installations (to address any concerns like visual impact of turbines or land use for solar farms) and demonstrate the environmental benefits to gain public support. Policy can encourage ports to use a portion of savings from efficiency gains in community projects (e.g., tree planting in the city, community renewable energy projects) to ensure the port-city relationship remains positive. This is more of a soft policy, but it is important for the social license to operate.

3.7.2 Phased Implementation Roadmap

Implementing these changes is a long-term process. We propose a phased roadmap consisting of four stages: Initiation, Expansion, Integration, and Continuous Governance. This roadmap is meant as a general guide; the timing and specifics might vary by port size and context, but the sequential approach helps manage complexity. This roadmap is designed as a generic, globally applicable framework, though its implementation details may vary by region depending on regulatory environments, infrastructure readiness, and economic capacity. For example, European ports may emphasize emissions compliance, while Indian and Chinese ports may prioritize capacity building and phased digitalization.

i. **Phase 1: Initiation (Pilot and Plan); Timeline: 1–2 years:**

   In this initial phase, the focus is on building the foundation and demonstrating quick wins. Key actions include: form the governance structures (e.g., Smart Port Committee), conduct comprehensive assessments (baseline study of current processes, energy use, and gaps in digital infrastructure), and identify pilot projects. One or two pilot projects combining IoT and energy goals should be launched to showcase potential benefits.

For example, a port might implement a pilot IoT yard management system in one terminal, or install a small solar-plus-battery unit to power an office building and a few charging stations. Another pilot could be a digital platform for trucking appointments to reduce gate congestion. During this phase, also develop the detailed master plan or strategy document for the smart green port, with input from stakeholders. Include targets (KPIs) and a high-level architecture. Importantly, secure initial funding – perhaps through budget allocation or grants for the pilots and capacity building. Early engagement with workforce and training on the pilot tech happens here, to start cultivating a culture of innovation. The goal of Initiation is to reduce scepticism by showing tangible improvements on a small scale and to build institutional momentum.

ii. **Phase 2: Expansion (Scale Up Infrastructure and Systems); Timeline: 2–5 years:**

In this phase, successful pilots are scaled up port-wide or to additional terminals, and more infrastructure is rolled out. For IoT, this might involve deploying port community system platforms connecting all stakeholders, equipping all major assets with sensors, rolling out port-wide connectivity (like a dedicated 5G network or LPWAN for IoT). For renewable energy, it's the period to install larger-scale projects: for example, covering all suitable roofs with solar panels, installing wind turbines if feasible, expanding the shore power infrastructure to multiple berths, procuring electric or hydrogen equipment to replace retiring diesel units. The expansion phase is typically capital-intensive and requires project management to minimize disruption to ongoing operations. It should be done in stages – for instance, retrofit one terminal with new tech while others operate normally, then rotate. During this phase, many processes go hybrid: some operations are digitized and optimized, while some legacy practices might still be in place as backup. Training programmes intensify to ensure staff can operate new systems at scale. Also, mid-phase, evaluate and fine-tune: data from initial implementations can identify bottlenecks (maybe the IoT network needs more bandwidth, or the first batch of electric trucks revealed the need for more charging stations than anticipated). Adjust investments accordingly. On the policy side, this is when some of the new regulations or standards previously planned may kick in – for instance, by year 3, mandate all data submission through the new system, by year 5, require that a certain percentage of yard equipment is electric, etc., aligning with the availability of infrastructure. By the end of Expansion, the port should have most of the necessary hardware and software in place, albeit some of it still operating in parallel with old systems.

iii. **Phase 3: Integration (Synergy and Full Implementation); Timeline: 5–7 years from start:**

In this phase, the various systems and infrastructure deployed are integrated to function as a cohesive smart port ecosystem. Operational processes are re-engineered to exploit the new capabilities. For example, the scheduling of ships, cranes, and trucks is now fully coordinated through the digital platform – this might involve eliminating older manual scheduling steps or legacy IT systems in favour of the unified one. The port's energy management system actively controls and optimizes usage – for instance, a central control can delay a non-urgent reefer container's defrost cycle to an off-peak time to shave load, or can choose optimal times to charge battery-electric vehicles based on solar output forecasts. The Physical, Digital, Decision layers we discussed should now be tightly linked: IoT sensors feed data to analytics that inform daily decisions by port managers, and governance protocols ensure each stakeholder acts on those insights. Phase 3 is also when interoperability across stakeholders is achieved ideally, by now shipping lines share their vessel ETA and cargo info digitally, truckers are integrated via mobile apps to the port scheduler, and terminal operators share performance data to a common dashboard. This level of integration often requires cultural adjustment the port community might need to trust the system's recommendations (e.g., an AI suggests an optimal stowage plan or truck turn time, and humans must embrace it). Continuous training and change management are important here. On the renewable side, integration may mean the port can operate as a microgrid if needed, testing an islanded mode to ensure resilience, or participating in demand-response with the main grid (selling back power at times, etc.). The port should measure and publicly report improvements: for example, X% reduction in average vessel stay, Y% energy now from renewables, Z% cut in CO_2emissions. Achieving these demonstrable outcomes is key for validation.

iv. **Phase 4: Governance and Optimization (Ongoing); Timeline: beyond 7 years:**

After initial integration, the focus shifts to continuous improvement and institutionalization of the new model. The port's governance structures should transition from planning mode to operational oversight mode. The Smart Port Committee might evolve into a permanent innovation council that oversees updates and ensures the port keeps up with technological advances (like adopting new AI models, or upgrading to 6G networks in the future, etc.). At this stage, the port could benchmark itself against other leading ports and collaborate internationally, participating in alliances like chainPORT or exchanging data via IAPH's data collaboration platforms to further optimize. Policy-wise, this is where enforcement of standards becomes routine, and adjustments can be made, for example, if one regulation is found too strict or too lenient, it can be calibrated. The port should also look outward: sharing its successes and frameworks with smaller ports in the country (if it's a hub port), effectively becoming a mentor. Many European ports do this via projects funded by EU, where a big port pilots and then rollout best practices to others. In this stage, the port can also integrate new trends: maybe autonomous ships start calling, requiring new protocols, or the hydrogen economy scales, requiring ports to serve as energy hubs. The governance should be flexible to incorporate these. Essentially, Phase 4 is about sustaining the smart and green practices, ensuring cybersecurity

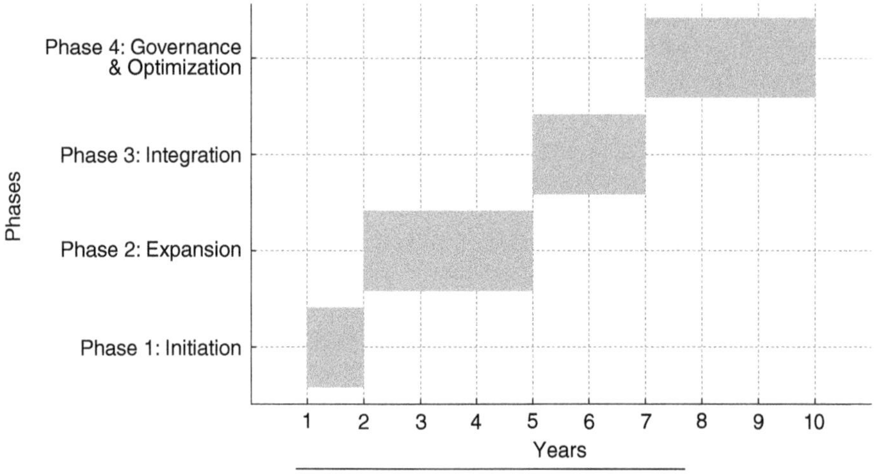

FIGURE 3.3 Four phases of the smart port implementation roadmap.

and system maintenance are ongoing priorities, and avoiding regression. It also involves updating the roadmap itself as new technologies (like maybe AI-driven predictive analytics or new energy storage tech) become viable; the port should incorporate continuous innovation.

Throughout all phases, monitoring and evaluation should be built in. Clear KPIs, such as truck turn time, crane productivity, energy cost per TEU, carbon emissions per TEU, etc. need to be tracked and reported. This not only shows progress but also highlights areas that need more attention (for instance, if data shows that one part of the operation isn't seeing improvement, targeted interventions can be made). The sequential progression of these four stages, along with their key milestones, is summarized in Figure 3.3.

3.7.3 International Benchmarking and Collaboration

An important recommendation is for ports (and countries) to leverage international collaboration. Many ports around the world are facing similar challenges and can learn from each other's experiences. For policymakers and port authorities, engaging in international forums such as the International Association of Ports and Harbours (IAPH) working groups on Data Collaboration and Climate & Energy, or the World Ports Climate Initiative can provide access to the latest best practices and tools. Benchmarking against global leaders like Rotterdam, Singapore, or Hamburg helps set realistic targets and identify proven solutions. For example, if Rotterdam achieved a particular reduction in truck idling through an AI truck management system, another port can take that as a benchmark and possibly adopt the same system or concept. Collaboration can also take the form of pilot partnerships: a developing country port might partner with a tech-savvy port and an international donor to pilot a solution similar to "twinning" arrangements. The IMO and World Bank have launched initiatives (like the Green Ports initiative under IMO's GloMEEP) to facilitate knowledge transfer ports should actively participate in these.

Finally, aligning port-level strategies with international regulations and agreements ensures longevity. For instance, if the IMO enforces stricter emissions measures for vessels by 2030, ports that already have shore power and LNG bunkering or hydrogen ready will be ahead of the curve in attracting business. International collaboration also extends to the supply chain: working with shipping lines and logistics providers on data integration creates a more resilient overall supply chain (as the pandemic disruptions taught, visibility across the chain is vital). Governments can support this by signing onto or initiating regional port agreements (e.g., a regional shore power deployment plan among neighbouring countries' ports so that ships have a consistent expectation).

In sum, the recommended policies and roadmap provide a comprehensive approach: incentivize and guide the needed investments, ensure people and institutions are prepared, enforce standards to keep everyone aligned, and phase the implementation to manage risk and allow learning. Through international cooperation and steadfast governance, ports can progressively turn the vision of IoT-driven, renewable-powered logistics into reality. The outcome will not only be decarbonized and efficient maritime supply chains, but also more competitive ports and healthier port-city environments.

3.8 CONCLUSION

The maritime industry is at a crossroads where digital transformation and sustainability imperatives converge. This chapter has explored how IoT-driven smart port logistics can be synergistically integrated with renewable energy systems to propel ports into a new era of efficiency and environmental responsibility. The integration potential is vast: IoT technologies can dramatically improve operational visibility and control in ports, leading to streamlined cargo flows, reduced vessel turnaround times, and predictive maintenance that minimizes downtime. When these digital capabilities are combined with renewable energy integration – such as ports generating solar or wind power on-site, electrifying equipment, and supplying clean shore power to ships – the result is a "smart port" that not only operates efficiently but also with a minimal carbon footprint. Case examples from leading ports illustrate that this is not a distant utopia but an emerging reality: for instance, Rotterdam's digital twin and AI systems optimize logistics while its investments in shore power and hydrogen aim to cut emissions, and Singapore's fully automated terminals coordinate closely with energy-efficient practices. In developing contexts like India, strides are being made through national initiatives to modernize ports with technology and greener infrastructure, though challenges remain in scaling these across diverse ports.

Several future research and development directions emerge from this study. One key area is the application of AI and predictive analytics on the rich data collected via IoT. As ports accumulate big data on ship movements, cargo patterns, equipment behaviour, and energy usage, advanced AI models (including machine learning and perhaps even neural networks) can be developed to forecast and optimize operations in ways humans cannot. For example, AI could predict congestion days in advance and suggest proactive measures, or dynamically optimize energy distribution in a port microgrid by learning from usage patterns. Research into custom AI algorithms for port logistics (like those that handle uncertainty in arrival times or disruptions) will enhance the "smartness" of ports. Another related frontier is the expansion of digital twin technology. Currently, only a few major ports have partial digital twins. In the future, more comprehensive digital twins – encompassing the entire port infrastructure, environment, and even parts of the supply chain – could be used for real-time monitoring and scenario analysis (for instance, simulating the impact of a new policy or a sudden weather event on port operations). Developing these twins requires interdisciplinary research (maritime engineering, computer science, data science) and could greatly aid port decision-makers by virtually testing options before physical implementation.

Secure data governance will also become increasingly important. As more data is shared among stakeholders, ports must ensure robust data protection and cybersecurity. Future work should explore frameworks for secure data sharing that balance openness and privacy – potentially leveraging technologies like blockchain for data integrity and access control in Port Community Systems (some have coined the term "Smart Port Blockchain" for secure logistics documentation). Research could evaluate the feasibility of blockchain or distributed ledger systems to facilitate trusted information exchange between shipping lines, customs, and ports, thereby reducing fraud and errors while maintaining privacy and security. Furthermore, with heightened cyber threats, ports will need to integrate AI in cybersecurity as well – such as AI systems that detect anomalies in network traffic of port OT (operational technology) systems indicative of a cyber intrusion.

From a governance perspective, an avenue for future study is how ports can incentivize broader supply chain decarbonization using their smart systems. For example, if a port's IoT platform extends to track hinterland transport (trucks/trains) emissions, ports could possibly reward shippers who choose greener hinterland options (like electric trucks or rail) by providing faster clearance or reduced fees. Thus, the port becomes an orchestrator of sustainability beyond its gates. This leads to the idea of ports as energy hubs in the future: as ports invest in renewables and perhaps energy storage, they could provide services not just to ships but to local grids (exporting power) and other industries (like providing hydrogen to nearby trucking or industrial clusters). Pilot projects in some European ports are looking into this, and further analysis is warranted on the economic and technical viability of ports serving as clean energy hubs for their regions.

In conclusion, IoT-driven smart port logistics and renewable energy integration are complementary strategies that together can transform maritime supply chains for the better. Smart ports employing IoT and AI can achieve unprecedented levels of efficiency, safety, and service reliability – critical for handling growing trade volumes and complex logistics. Concurrently, by embracing renewables and electrification, ports significantly reduce their environmental impact and help meet global climate targets, all while potentially lowering long-term energy costs and improving resilience (through microgrids and reduced exposure to fuel price volatility). This chapter's integrated framework and recommendations provide a roadmap for stakeholders to navigate this transformation responsibly and effectively. Achieving this vision will require overcoming the technical, financial, and institutional barriers discussed, but the trend is clearly towards smarter and greener ports.

Ports have always been pivotal nodes in global commerce; in the coming decades, they are poised to become exemplars of industrial innovation and sustainability. By harnessing IoT and clean energy, the "ports of the future" will enable not just faster and cheaper trade, but also cleaner and smarter trade. Continued collaboration among port authorities, industry, governments, and the research community will be essential to drive innovation, share best practices, and ensure that even smaller or less-developed ports are not left behind in this transition. Future research should continue to address the gaps and emerging challenges – from cyber-physical security to economic modelling of green investments – to support evidence-based decision-making. Ultimately, the journey to smart, sustainable ports is a marathon, not a sprint, but it is underway and accelerating. Ports that proactively integrate digital and green strategies will lead the way in a decarbonized, intelligent global supply chain, aligning economic efficiency with environmental stewardship for generations to come.

REFERENCES

Abu Bakar, N. N., Bazmohammadi, N., Vasquez, J. C., & Guerrero, J. M. (2023a). Electrification of onshore power systems in maritime transportation towards decarbonization of ports: A review of the cold ironing technology. *Renewable and Sustainable Energy Reviews*, *178*, 113243. https://doi.org/10.1016/J.RSER.2023.113243

Abu Bakar, N. N., Bazmohammadi, N., Vasquez, J. C., & Guerrero, J. M. (2023b). Electrification of onshore power systems in maritime transportation towards decarbonization of ports: A review of the cold ironing technology. *Renewable and Sustainable Energy Reviews*, *178*, 113243. https://doi.org/10.1016/J.RSER.2023.113243

Alahmadi, D. H., Baothman, F. A., Alrajhi, M. M., Alshahrani, F. S., & Albalawi, H. Z. (2021). Comparative analysis of blockchain technology to support digital transformation in ports and shipping. *Journal of Intelligent Systems*, *31*(1), 55–69. https://doi.org/10.1515/jisys-2021-0131

Alam, M., Dupras, J., & Messier, C. (2016). A framework towards a composite indicator for urban ecosystem services. *Ecological Indicators*, *60*, 38–44. https://doi.org/10.1016/J.ECOLIND.2015.05.035

Alamoush, A. S., Ballini, F., & Ölçer, A. I. (2021). Revisiting port sustainability as a foundation for the implementation of the United Nations Sustainable Development Goals (UN SDGs). *Journal of Shipping and Trade*, *6*(1), 19. https://doi.org/10.1186/s41072-021-00101-6

Aslam, S., Navarro, A., Aristotelous, A., Garro Crevillen, E., Martınez-Romero, A., Martínez-Ceballos, Á., Cassera, A., Orphanides, K., Herodotou, H., & Michaelides, M. P. (2025). Machine learning-based predictive maintenance at smart ports using IoT sensor data. *Sensors*, *25*(13), 3923. https://doi.org/10.3390/s25133923

Barberi, S., Campisi, T., & Neduzha, L. (2022). The role of cold ironing in maritime transport emissions. *AIP Conference Proceedings*, *2611*. https://doi.org/10.1063/5.0119881

Belmoukari, B., Audy, J.-F., & Forget, P. (2023). Smart port: A systematic literature review. *European Transport Research Review*, *15*(1), 4. https://doi.org/10.1186/s12544-023-00581-6

Bocayuva, M. (2021). Cybersecurity in the European Union port sector in light of the digital transformation and the COVID-19 pandemic. *WMU Journal of Maritime Affairs*, *20*(2), 173–192. https://doi.org/10.1007/s13437-021-00240-4

Bolbot, V., Kulkarni, K., Brunou, P., Banda, O. V., & Musharraf, M. (2022). Developments and research directions in maritime cybersecurity: A systematic literature review and bibliometric analysis. *International Journal of Critical Infrastructure Protection, 39*, 100571. https://doi.org/10.1016/J.IJCIP.2022.100571

Cavalli, L., Lizzi, G., Guerrieri, L., Querci, A., De Bari, F., Barbieri, G., Ferrini, S., Di Meglio, R., Cardone, R., Tardo, A., Pagano, P., Tesei, A., & Lattuca, D. (2021). Addressing efficiency and sustainability in the port of the future with 5g: The experience of the Livorno port. A methodological insight to measure innovation technologies' benefits on port operations. *Sustainability (Switzerland), 13*(21), 12146. https://doi.org/10.3390/su132112146

Chen, J., Huang, T., Xie, X., Lee, P. T.-W., & Hua, C. (2019). Constructing governance framework of a green and smart port. *Journal of Marine Science and Engineering*, *7*(4), 83. https://doi.org/10.3390/jmse7040083

Dinh, G. H., Pham, H. T., Nguyen, L. C., Dang, H. Q., & Pham, N. D. K. (2024). Leveraging artificial intelligence to enhance port operation efficiency. *Polish Maritime Research, 31*(2), 140–155. https://doi.org/10.2478/pomr-2024-0030

Ducruet, C., Polo Martin, B., Sene, M. A., Lo Prete, M., Sun, L., Itoh, H., & Pigné, Y. (2024). Ports and their influence on local air pollution and public health: A global analysis. *Science of the Total Environment, 915*, 170099. https://doi.org/10.1016/J.SCITOTENV.2024.170099

Dzemydienė, D., Burinskienė, A., Čižiūnienė, K., & Miliauskas, A. (2023). Development of E-service provision system architecture based on IoT and WSNs for monitoring and management of freight intermodal transportation. *Sensors*, *23*(5), 2831. https://doi.org/10.3390/s23052831

Elkafas, A. G., & Seddiek, I. S. (2024). Application of renewable energy systems in seaports towards sustainability and decarbonization: Energy, environmental and economic assessment. *Renewable Energy*, *228*, 120690. https://doi.org/10.1016/J.RENENE.2024.120690

Farah, M. Ben, Ahmed, Y., Mahmoud, H., Shah, S. A., Al-kadri, M. O., Taramonli, S., Bellekens, X., Abozariba, R., Idrissi, M., & Aneiba, A. (2024). A survey on blockchain technology in the maritime industry: Challenges and future perspectives. *Future Generation Computer Systems*, *157*, 618–637. https://doi.org/10.1016/J.FUTURE.2024.03.046

Ferrari, E., Christidis, P., & Bolsi, P. (2023). The impact of rising maritime transport costs on international trade: Estimation using a multi-region general equilibrium model. *Transportation Research Interdisciplinary Perspectives*, *22*, 100985. https://doi.org/10.1016/J.TRIP.2023.100985

Gerakoudi, K., Kokosalakis, G., & Stavroulakis, P. J. (2024). A machine learning approach towards reviewing the role of 'Internet of Things' in the shipping industry. *Journal of Shipping and Trade*, *9*(1), 19. https://doi.org/10.1186/s41072-024-00177-w

González-Cancelas, N., Camarero Orive, A., Vilarchao, A. R., & Vaca-Cabrero, J. (2024). Use of End-to-end tool for the analysis of the digital governance of ports. *Logistics*, *8*(2), 58. https://doi.org/10.3390/logistics8020058

Heikkilä, M., Saarni, J., & Saurama, A. (2022). Innovation in smart ports: Future directions of digitalization in container ports. *Journal of Marine Science and Engineering*, *10*(12), 1925. https://doi.org/10.3390/jmse10121925

Hentschel, M., Ketter, W., & Collins, J. (2018). Renewable energy cooperatives: Facilitating the energy transition at the Port of Rotterdam. *Energy Policy*, *121*, 61–69. https://doi.org/10.1016/J.ENPOL.2018.06.014

Inkinen, T., Helminen, R., & Saarikoski, J. (2019). Port digitalization with open data: Challenges, opportunities, and integrations. *Journal of Open Innovation: Technology, Market, and Complexity*, *5*(2), 30. https://doi.org/10.3390/JOITMC5020030

Inkinen, T., Helminen, R., & Saarikoski, J. (2021). Technological trajectories and scenarios in seaport digitalization. *Research in Transportation Business & Management*, *41*, 100633. https://doi.org/10.1016/J.RTBM.2021.100633

Innes, A., & Monios, J. (2018). Identifying the unique challenges of installing cold ironing at small and medium ports – The case of Aberdeen. *Transportation Research Part D: Transport and Environment*, *62*, 298–313. https://doi.org/10.1016/J.TRD.2018.02.004

Inutsuka, H., Ichimura, K., Sugimura, Y., Yoshie, M., & Shinoda, T. (2024). Study on the relationship between port governance and terminal operation system for smart port: Japan case. *Logistics*, *8*(2), 59. https://doi.org/10.3390/logistics8020059

Iris, Ç., & Lam, J. S. L. (2021). Optimal energy management and operations planning in seaports with smart grid while harnessing renewable energy under uncertainty. *Omega*, *103*, 102445. https://doi.org/10.1016/J.OMEGA.2021.102445

Jin, J., & Guo, Y. (2025). Exploring the drivers of digital transformation in Chinese port and shipping enterprises: A machine learning approach. *PLOS ONE*, *20*(5), 1–15. https://doi.org/10.1371/journal.pone.0322872

Jović, M., Tijan, E., Brčić, D., & Pucihar, A. (2022). Digitalization in maritime transport and seaports: Bibliometric, content and thematic analysis. *Journal of Marine Science and Engineering*, 10(4), 486. MDPI. https://doi.org/10.3390/jmse10040486

Klar, R., Fredriksson, A., & Angelakis, V. (2023). Digital twins for ports: Derived from smart city and supply chain twinning experience. *IEEE Access*, *11*, 71777–71799. https://doi.org/10.1109/ACCESS.2023.3295495

Lacalle, I., Belsa, A., Vaño, R., & Palau, C. E. (2020). Framework and methodology for establishing port-city policies based on real-time composite indicators and IoT: A practical use-case. *Sensors*, *20*(15), 4131. https://doi.org/10.3390/s20154131

Le, S.-T. (2024). Research on drivers and barriers to the implementation of cold ironing technology in zero emissions port. *Environmental Health Insights*, *18*, 11786302241265090. https://doi.org/10.1177/11786302241265090

Li, K., Gharehgozli, A., & Lee, J. Y. (2023). Smart technologies and port operations: Optimal adoption strategy with network externality consideration. *Computers & Industrial Engineering*, *184*, 109557. https://doi.org/10.1016/J.CIE.2023.109557

Liu, J., Zhang, H., & Zhen, L. (2023). Blockchain technology in maritime supply chains: Applications, architecture and challenges. *International Journal of Production Research*, *61*(11), 3547–3563. https://doi.org/10.1080/00207543.2021.1930239

Molavi, A., Lim, G. J., & Race, B. (2020). A framework for building a smart port and smart port index. *International Journal of Sustainable Transportation*, *14*(9), 686–700. https://doi.org/10.1080/15568318.2019.1610919

Moros-Daza, A., Amaya-Mier, R., & Paternina-Arboleda, C. (2020). Port community systems: A structured literature review. *Transportation Research Part A: Policy and Practice*, *133*, 27–46. https://doi.org/10.1016/J.TRA.2019.12.021

Mukherjee, A., Sasidharan, M., Herrera, M., & Parlikad, A. K. (2024). Unsupervised constrained discord detection in IoT-based online crane monitoring. *Advanced Engineering Informatics*, *60*, 102444. https://doi.org/10.1016/J.AEI.2024.102444

Muñuzuri, J., Onieva, L., Cortés, P., & Guadix, J. (2020). Using IoT data and applications to improve port-based intermodal supply chains. *Computers & Industrial Engineering*, *139*, 105668. https://doi.org/10.1016/J.CIE.2019.01.042

Ouyang, R., Wang, J., Xu, H., Chen, S., Xiong, X., Tolba, A., & Zhang, X. (2023). A Microservice and serverless architecture for secure IoT system. *Sensors*, *23*(10), 4868. https://doi.org/10.3390/s23104868

Pham, T. Y. (2023). A smart port development: Systematic literature and bibliometric analysis. *The Asian Journal of Shipping and Logistics*, *39*(3), 57–62. https://doi.org/10.1016/J.AJSL.2023.06.005

Philipp, R., Prause, G., Olaniyi, E. O., & Lemke, F. (2021). Towards green and smart seaports: Renewable energy and automation technologies for bulk cargo loading operations. *Environmental and Climate Technologies*, *25*(1), 650–665. https://doi.org/10.2478/rtuect-2021-0049

Prousaloglou, P., Kyriakopoulou-Roussou, M. C., Stavroulakis, P. J., Tsioumas, V., & Papadimitriou, S. (2025). Artificial intelligence in the service of sustainable shipping. *Journal of Ocean Engineering and Marine Energy*, *11*(3), 621–653. https://doi.org/10.1007/s40722-025-00390-0

Sifakis, N., Konidakis, S., & Tsoutsos, T. (2021). Hybrid renewable energy system optimum design and smart dispatch for nearly Zero Energy Ports. *Journal of Cleaner Production*, *310*, 127397. https://doi.org/10.1016/J.JCLEPRO.2021.127397

Surucu-Balci, E., Iris, Ç., & Balci, G. (2024). Digital information in maritime supply chains with blockchain and cloud platforms: Supply chain capabilities, barriers, and research opportunities. *Technological Forecasting and Social Change*, *198*, 122978. https://doi.org/10.1016/J.TECHFORE.2023.122978

Tawfik, M., Shehata, A. S., Hassan, A. A., & Kotb, M. A. (2023). Renewable solar and wind energies on buildings for green ports in Egypt. *Environmental Science and Pollution Research*, *30*(16), 47602–47629. https://doi.org/10.1007/s11356-023-25403-z

Tijan, E., Aksentijević, S., Ivanić, K., & Jardas, M. (2019). Blockchain technology implementation in logistics. *Sustainability*, *11*(4), 1185. https://doi.org/10.3390/su11041185

Wang, L., & Hsu, H.-H. (2025). IoT technology in maritime logistics management: exploration of data analysis methods. *Discover Internet of Things*, *5*(1), 66. https://doi.org/10.1007/s43926-025-00167-9

Yang, Y., Zhong, M., Yao, H., Yu, F., Fu, X., & Postolache, O. (2018). Internet of things for smart ports: Technologies and challenges. *IEEE Instrumentation & Measurement Magazine*, *21*(1), 34–43.

Yap, W. Y., Hsieh, C. H., & Lee, P. T. W. (2023). Shipping connectivity data analytics: Implications for maritime policy. *Transport Policy*, *132*, 112–127. https://doi.org/10.1016/J.TRANPOL.2022.12.018

Yigit, Y., Nguyen, L. D., Ozdem, M., Kinaci, O. K., Hoang, T., Canberk, B., & Duong, T. Q. (2023). TwinPort: 5G drone-assisted data collection with digital twin for smart seaports. *Scientific Reports*, *13*(1), 12310. https://doi.org/10.1038/s41598-023-39366-1

Yu, P., Zhaoyu, W., Yifen, G., Nengling, T., & Jun, W. (2023). Application prospect and key technologies of digital twin technology in the integrated port energy system. *Frontiers in Energy Research*, *10*, 1044978. https://doi.org/10.3389/fenrg.2022.1044978

Zaychenko, I., Smirnova, A., Gorshechnikova, P., & Piminov, N. (2021). Sustainable digital transformation of the port equipment management system. *E3S Web of Conferences*, *258*. EDP Sciences. https://doi.org/10.1051/e3sconf/202125802014

Zerbino, P., Aloini, D., Dulmin, R., & Mininno, V. (2019). Towards analytics-enabled efficiency improvements in maritime transportation: A case study in a mediterranean port. *Sustainability (Switzerland)*, *11*(16), 4473. https://doi.org/10.3390/su11164473

4 Technological Innovations in Solar Energy and Their Impact on Environmental, Social, and Governance Goals

Jay Patel and Neeta Patel

NOMENCLATURE

Symbol	Description
Abbreviation	
AI	Artificial Intelligence
ANN	Artificial Neural Networks
CAGR	Compound Annual Growth Rate
CSP	Concentrated Solar Power
EPR	Extended Producer Responsibility
ESG	Environmental, Social, and Governance
FPV	Floating Solar Photovoltaic
IRENA	International Renewable Energy Agency
IoT	Internet of Things
LSTM	Long Short-Term Memory
PEC	Power Efficiencies Conversion
PPP	Public-Private Partnerships
PSC	Perovskite Solar Cell
PV	Photovoltaic
TES	Thermal Energy Storage
UNSDG	United Nation Sustainable Development Goals

4.1 INTRODUCTION

The twenty-first century has witnessed a significant paradigm shift in the global energy landscape, driven by the urgent need to transition towards sustainable and low-carbon energy sources. Among various renewable energy technologies *'solar energy'* has emerged as one of the most promising and rapidly advancing solutions capable of addressing Environmental, Social, and Governance (ESG) challenges. Solar energy

DOI: 10.1201/9781003671039-4

is at the forefront of international efforts to reduce climate change, improve energy security, and promote inclusive socioeconomic development because of its widespread availability, scalability, and falling cost structure. As the global community intensifies efforts to meet the UNSDGs by 2030, innovations in solar technologies are playing a pivotal role in shaping a sustainable and resilient future (Patel, 2025).

Recent advancements in solar energy technologies have redefined the efficiency and applicability of solar power systems, offering breakthroughs that extend beyond conventional photovoltaic (PV) modules. Figure 4.1 shows that emerging technologies such as perovskite solar cells, tandem solar cells, and bifacial panels have pushed conversion efficiencies to record-breaking levels, surpassing the traditional Shockley-Queisser limit of 29–30% for single-junction silicon solar cells (NREL, 2023a). For instance, tandem solar cells, which combine perovskite and silicon layers, have achieved certified efficiencies above 32% in laboratory conditions, representing a significant leap over traditional silicon-based PV modules, which plateaued around 22–24% in commercial deployment (Nagaraja et al., 2025). Such enhancements not only improve the energy yield but also reduce the overall material intensity and life cycle carbon footprint of solar systems, thus directly contributing to the *'Environmental'* aspect of ESG.

Furthermore, innovations in Concentrated Solar Power (CSP) systems, particularly those systems which are integrated with Thermal Energy Storage (TES), have substantially enhanced the dispatch ability of solar energy (Patel & Patel, 2024b). CSP systems, utilizing molten salt storage or other advanced thermal storage media, enable continuous energy generation well beyond sunlight hours to ensure grid stability and reliability. Meanwhile the critical need of CSP plants is to deliver electricity during peak demand periods, aligning with the critical need for flexible and resilient energy infrastructure in a low-carbon economy (Patel & Patel, 2023). There is a compelling argument for replacing baseload fossil power with renewable alternatives since CSP plants, which have 10–15 hours of thermal storage, have shown capacity factors above 50%, which is comparable to traditional fossil fuel plants.

Solar power optimization has been further transformed by Artificial Intelligence (AI) and solar ray tracing technologies. Solar ray tracing simulations employ detailed geometrical and atmospheric models to accurately predict solar irradiance over complex terrains and built environments. This precise prediction enables the optimal placement and orientation of solar panels and reflectors and that helps in maximizing energy capture and system performance. AI-driven predictive maintenance also reduces operational costs and extends system lifetimes, reinforcing the governance principle of operational efficiency and transparency. Solar technologies have shown significant societal impacts in the search for decentralized energy alternatives. Agrivoltaics, which involves the co-location of agriculture and solar PV panels, provides a unique opportunity to enhance land use efficiency, improve crop yields through microclimatic modifications, and increase farm incomes (Supe et al., 2024). Studies have shown that agrivoltaics systems can enhance overall land productivity by 60–70% compared to monoculture agriculture or standalone solar fields. Similarly, floating solar farms, deployed over reservoirs and lakes, address land scarcity issues and reduce evaporation losses, offering integrated water-energy solutions critical in water-stressed regions.

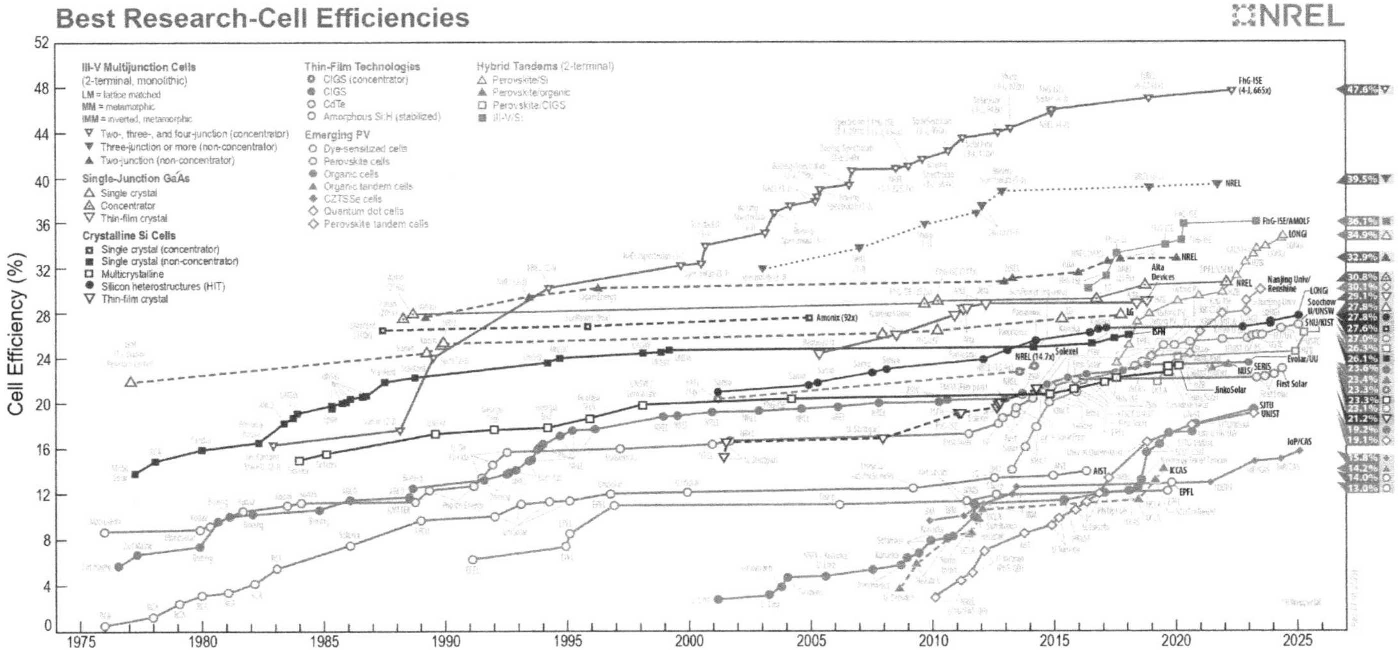

FIGURE 4.1 The highest confirmed conversion efficiencies for research cells for a range of photovoltaic technologies, plotted from 1976 to the present (NREL, 2023b).

Furthermore, innovations and technological improvements in solar microgrids have strengthened their role in reducing global carbon emissions. In addition, solar microgrids have become a cornerstone for rural electrification and provide reliable electricity access to over 100 million people, primarily in Sub-Saharan Africa and South Asia (Jignesh et al., 2018). Access to solar-powered electricity has been directly linked to improvements in educational outcomes, healthcare delivery (through reliable refrigeration for vaccines), and the creation of local economic opportunities. On the other hand, financially the solar energy sector has also witnessed unprecedented growth, with global investments in solar projects. ESG-driven investment vehicles, including green bonds and climate funds, have significantly contributed to this momentum. Green bonds earmarked for renewable energy projects have grown at a Compound Annual Growth Rate (CAGR) of 25% over the past five years, signalling strong investor confidence in sustainable energy assets (Steffen, 2018). Furthermore, the integration of carbon credit mechanisms has incentivized corporate actors to invest in solar initiatives as part of their decarbonization strategies, aligning financial incentives with environmental objectives. Public-Private Partnerships (PPPs) have played a critical role in scaling solar energy infrastructure, particularly in emerging economies where public sector financing alone is insufficient (Patel & Singh, 2015b). Successful PPP models have demonstrated that risk-sharing mechanisms and blended finance structures can attract private capital while ensuring that social and environmental safeguards are maintained. For instance, India's solar park programme under the PPP model has added over 10 GW of solar capacity while creating more than 100,000 jobs which highlights the synergistic potential of collaborative governance frameworks (Raghutla & Kolati, 2023).

Despite the substantial progress in solar technologies, several challenges persist that threaten to hinder the full realization of solar energy's ESG potential. Supply chain sustainability, especially the ethical sourcing of critical raw materials like polysilicon, cobalt, and rare earth elements, remains a major concern (Priyanshu et al., 2018; Vadhera et al., 2018). The energy-intensive production processes associated with these materials, coupled with labour rights violations in certain supply regions, necessitate more stringent supply chain transparency and responsible sourcing initiatives. Furthermore, managing solar panel end-of-life presents a serious environmental issue; projections indicate that cumulative solar waste worldwide may surpass 78 million metric tonnes by 2050 (Patel et al., 2023). Thus, advancing circular economy approaches through material recycling and panel repurposing is imperative. Equitable access to solar financing is another pressing issue. High upfront capital costs, lack of credit access, and inadequate policy support continue to hinder solar adoption among low-income communities and small enterprises (Patel et al., 2024).

Overall, solar technological innovations offer transformative pathways for advancing the ESG agenda and accelerating the global transition to a sustainable, low-carbon economy. A strong basis for sustainable development has been established by the combination of decentralized energy systems, AI-driven optimization, high-efficiency PV technologies, and cutting-edge financial instruments. The integration of Internet of Things (IoT) as infrastructure enables optimization of solar applications which enhance operational efficiency and reliability. This connectivity-driven approach strengthens the data-driven decision-making and supports adaptive energy management. However, addressing supply chain vulnerabilities, enhancing circularity, and ensuring inclusive financing are critical to fully harnessing solar energy's potential for long-term global sustainability. The present study aims to explore these dimensions in depth by analysing the latest technological breakthroughs in solar energy, evaluating their alignment with ESG principles, and assessing their contributions to the UNSDGs. With this discussion it is observed that solar innovations from the perspectives of engineering, finance, and governance help in shaping the future of sustainable energy systems in the long run.

4.2 REVOLUTIONS IN SOLAR TECHNOLOGIES

4.2.1 Solar PV Technologies

PV technology has undergone significant evolution since its inception. The first-generation PV cells, primarily silicon-based, have dominated the market due to their established manufacturing processes and relatively high efficiencies. From literature it is observed that the annual solar PV cell production capacity of different countries from the year 2010 to 2021. Like, China has played an outsized role in scaling up the mass production of solar PV cells and modules, comprising 78% of global production in 2021. However, noticeably next big player is Malaysia and rest of world (ROW) is similar to Malaysia in numbers (Helveston et al., 2022). Moreover, from comparison point of view a projected annual cost savings from 2020 to 2030 for China, Germany, and the United States under two scenarios, (a) global market and (b) national market conditions show the following scenarios. The savings are estimated by multiplying the installed solar capacity of each country each year by the difference between the modelled prices in these two scenarios. The error bars indicate 95% confidence intervals, calculated through simulation. However, their production involves energy-intensive processes and higher material costs. To address these limitations, second-generation thin-film technologies, such as Cadmium Telluride (CdTe) and Copper Indium Gallium Selenide (CIGS) were developed (Kim et al., 2014; Venkatesh et al., 2024). These offer reduced material usage and manufacturing costs but often at the expense of lower efficiencies. The third generation of PV technologies introduces materials like perovskites and quantum dots, aiming to combine high efficiency with low production costs. In the following section the different PV technologies and their impact on ESG have been discussed.

4.2.1.1 Perovskite Solar Cells

Perovskite Solar Cells (PSCs) have emerged as a promising third-generation PV technology due to their exceptional optoelectronic properties, including high absorption coefficients and tuneable bandgaps. Laboratory-scale PSCs have achieved Power Conversion Efficiencies (PCEs) exceeding 25%, with tandem configurations surpassing 30%. The rapid efficiency improvements, coupled with low fabrication costs, make PSCs attractive for commercial applications (Afroz et al., 2025). However, challenges such as long-term stability, scalability, and the use of lead-based materials remain areas of active research.

4.2.1.2 Tandem and Multi-junction Cells

Tandem and multi-junction solar cells aim to overcome the efficiency limitations of single-junction devices by stacking multiple layers with different bandgaps, allowing for broader spectral absorption. Silicon-perovskite tandem cells have demonstrated significant efficiency enhancements, with recent developments achieving certified efficiencies of 27.1% (Hussain & Gagliardi, 2022). These configurations leverage the high efficiency of perovskite top cells and the stability of silicon bottom cells. Further advancements in triple-junction cells and incorporating additional perovskite layers have the potential to exceed 30% efficiency which offers a pathway to surpass the theoretical limits of single-junction cells (Yao et al., 2023).

4.2.1.3 Bifacial Solar Panels

Bifacial solar panels represent an innovative approach to enhancing energy yield by capturing sunlight from both the front and rear surfaces. These panels can achieve up to 15–20% more energy output compared to traditional mono-facial panels, particularly in high-albedo environments where reflected light is abundant (Sahu et al., 2023). The adoption of bifacial technology is facilitated by advancements in module design and mounting systems that optimize rear-side exposure. However, accurate performance modelling and standardized testing protocols are essential to fully realize the benefits of bifacial systems in various deployment scenarios.

Overall, advancements in solar PV technologies hold profound implications for ESG dimensions. From an environmental perspective, the enhanced efficiency of next-generation PV systems reduces the spatial footprint required for solar installations, thereby minimizing habitat disruption and mitigating land use conflicts. In parallel, the transition towards materials with lower environmental impact along with innovations in end-of-life recycling of PV modules supports a more sustainable and circular life cycle management framework. From a social standpoint, the widespread deployment of decentralized solar technologies including bifacial modules and tandem cells has significantly improved energy accessibility in remote and underserved regions. This democratization of clean energy contributes to community empowerment, strengthens local energy security, and fosters socioeconomic development by facilitating distributed power generation and creating employment opportunities. From a governance perspective, the integration of advanced solar technologies into national and regional energy infrastructures necessitates the formulation of comprehensive policy frameworks and regulatory standards that ensure quality assurance, operational safety, and long-term reliability (Lorenzo, 2021). Furthermore, transparent ESG reporting protocols and standardized performance metrics are critical to monitoring sustainability outcomes as well as enhancing investor trust in solar-based infrastructure projects.

4.2.2 CSP and Thermal Storage

CSP systems harness solar energy by employing reflective devices such as mirrors or lenses, to concentrate sunlight onto a specific receiver. This concentrated solar radiation is then converted into thermal energy which can be utilized to produce electricity through conventional thermodynamic cycles. CSP technologies are categorized into four primary types: parabolic troughs, solar power towers, linear Fresnel reflectors, and dish Stirling systems (Kumar et al., 2017). Each of these configurations focuses sunlight to heat up a working fluid, typically a molten salt or water steam, which then drives a turbine connected to an electrical generator (Patel & Patel, 2023). The efficiency of CSP systems is influenced by factors such as the concentration ratio, thermal losses, and the thermodynamic efficiency of the power cycle employed.

4.2.2.1 Integration with Thermal Energy Storage

A significant advancement in CSP technology is the integration of TES systems particularly those utilizing molten salts. These TES systems enable CSP plants to store excess thermal energy during periods of high solar irradiance and dispatch electricity during periods of low or no sunlight, thereby enhancing the dispatch ability and reliability of solar power (Patel & Singh, 2015a). Molten salt storage systems can maintain thermal energy at high temperatures, typically between 290°C and 565°C. Notable implementations include the Crescent Dunes Solar Energy Project in Nevada, which features a 10-hour molten salt storage capacity, and the Solana Generating Station in Arizona, equipped with a 6-hour storage system (Patel, 2025). These integrations have demonstrated the feasibility of CSP plants to provide continuous power supply, aligning with grid demands and reducing reliance on fossil fuel-based peaking power plants.

4.2.2.2 Environmental and Grid Benefits

The deployment of CSP systems with integrated TES offers substantial environmental benefits by facilitating the reduction of greenhouse gas emissions and mitigating climate change impacts. By providing a stable and dispatchable renewable energy source, CSP plants contribute to grid stability and reliability (Cellura et al., 2025). The ability of CSP systems to supply power during peak demand periods reduces the need for fossil fuel-based peaking plants that helps in decreasing carbon emissions and enhancing the sustainability of the energy sector. Furthermore, CSP plants can be strategically integrated into existing power grids, offering ancillary services such as frequency regulation and voltage support, which are essential for maintaining grid stability in the presence of variable renewable energy sources (Costa et al., 2024).

4.2.2.3 Socioeconomic Advantages

Beyond environmental benefits, CSP projects have significant socioeconomic impacts, particularly in terms of job creation and skill development. The construction, operation, and maintenance of CSP plants require a diverse workforce, leading to employment opportunities in engineering, construction, and technical services. For instance, during Spain's CSP deployment period from 2007 to 2013, approximately 30,000 jobs were generated, underscoring the technology's potential for economic development (Corona et al., 2016). Additionally, CSP projects often involve local communities, fostering technological skill development and capacity-building. This engagement not only supports the successful implementation and operation of CSP plants but also contributes to the broader goal of sustainable development by enhancing local expertise and promoting energy independence.

4.2.3 AI-Integrated IoT Infrastructure for Smart Solar Energy Solutions

4.2.3.1 Solar Ray Tracing Algorithms

Advanced solar ray tracing algorithms are pivotal in optimizing PV as well as solar thermal systems performance by accurately modelling the interaction between the available solar radiation and the system (Patel & Patel, 2024a). These simulations consider variables such as sun position, panel orientation, shading effects, and atmospheric conditions to enhance irradiance capture. For instance, a study by the National Renewable Energy Laboratory (NREL) demonstrated that implementing ray tracing techniques in PV system design can improve annual energy yield by up to 15% compared to traditional methods (NREL, 2023b). Moreover, integrating ray tracing with optimization algorithms enables the development of dynamic tracking systems. These systems adjust panel angles in real time to maximize solar exposure, leading to an additional 20–25% increase in energy production over fixed-tilt systems. Such advancements not only boost energy efficiency but also contribute to reducing the cost of electricity from solar installations.

4.2.3.2 AI-Integrated IoT-Based Smart Solar Infrastructure

AI-integrated IoT-based smart solar infrastructure represents a transformative approach to renewable energy utilization. By embedding IoT sensors within solar panels and energy storage units, real-time data on irradiance, temperature, and load demand can be continuously monitored. AI algorithms process this data to optimize energy generation, distribution, and storage with minimal human intervention. Moreover, predictive models enable accurate forecasting of solar output, thereby improving system reliability under variable weather conditions (Mamodiya et al., 2025). Integration with intelligent demand-side management ensures efficient utilization of solar power for both residential and industrial consumers. Such systems can dynamically switch between grid-connected and off-grid modes, depending on energy availability and user requirements.

Additionally, AI has emerged as a transformative tool in forecasting energy generation and its optimum performance based on solar radiation availability. Machine learning models, including Artificial Neural Networks (ANN), support vector machines, and Long Short-Term Memory (LSTM) networks, have been employed to predict solar power output with remarkable accuracy. Machine learning techniques can identify performance degradation in solar panels, reducing maintenance costs through predictive diagnostics. IoT-enabled communication further facilitates interoperability among multiple solar units, creating scalable smart microgrids. The infrastructure enhances energy resilience in regions with unstable power supply by ensuring continuous operation through optimized storage control. Advanced AI-driven control strategies have the potential to achieve efficiency improvements of 15–25% compared to conventional solar systems. In addition, this technology contributes to reducing carbon emissions, thereby supporting global sustainability and climate action goals (Yu et al., 2024a). Overall, AI-integrated IoT-based smart solar infrastructure offers a pathway towards intelligent, adaptive, and cost-effective renewable energy systems.

A comparative study conducted in Morocco evaluated six machine learning algorithms for solar energy forecasting and found that the ANN model achieved the highest performance, with a coefficient of determination of 0.98 and a root mean squared error of 0.15 (Ledmaoui et al., 2023). Further advancements involve hybrid models that combine different AI techniques to enhance prediction accuracy. Like, a study integrating singular spectrum analysis, convolutional neural networks, and LSTM networks achieved a mean absolute error of 0.12 for one-hour-ahead solar power forecasts in greenhouse applications (Venkateswaran & Cho, 2024). These high-precision forecasts are crucial for grid operators to manage supply-demand balance, reduce curtailment, and integrate higher shares of renewable energy into the grid. Overall, the synergy between advanced solar ray tracing algorithms and AI-based energy forecasting not only optimizes technical performance but also reinforces ESG objectives. By harnessing these technologies, stakeholders can drive the adoption of sustainable energy solutions that are efficient, equitable, and transparent.

4.2.4 Decentralized and Innovative Solar Applications

Decentralized and innovative solar applications such as agrivoltaics (Agriculture-integrated solar systems), Floating solar farms, and Off-grid solar microgrids represent emerging strategies for harnessing solar energy through dual or underutilized land use. These technologies aim to optimize land productivity by utilizing agricultural fields, water bodies, and remote regions for sustainable energy generation. The following sections provide a detailed discussion of each of these applications and their contributions to environmental and social sustainability.

4.2.4.1 Agrivoltaics

Agrivoltaics, the synergistic integration of PV systems with agricultural practices, presents a dual-use approach to land management, optimizing both energy production and crop cultivation. Empirical studies have demonstrated that agrivoltaic systems can enhance land productivity by up to 60% compared to traditional monoculture or standalone solar installations (Garrod et al., 2024). This improvement is primarily attributed to the microclimatic advantages offered by solar panels, such as partial shading, which reduces soil moisture loss through evaporation and alleviates heat stress on crops. For instance, research conducted in arid regions like the Sonoran desert in Arizona has shown that crops grown under solar panels exhibit improved resilience and yield due to moderated temperatures and reduced water stress (Majumdar & Pasqualetti, 2018). Despite these advantages the adoption of agrivoltaics remains limited and accounts for only about 1% of global solar power generation. Challenges such as increased initial costs and the need for specialized infrastructure have hindered widespread implementation. Nevertheless, with growing concerns over land use efficiency and sustainable agriculture, agrivoltaics offers a promising pathway towards integrated food and energy systems.

4.2.4.2 Floating Solar Farms

Floating Solar Photovoltaic (FPV) systems, installed on bodies of water like reservoirs and lakes, offer a novel solution to land scarcity and water conservation challenges. These systems not only generate renewable energy but also significantly reduce water evaporation and work to harness the available solar energy on the water surfaces. According to research on the Aswan High Dam reservoir in Egypt, 90% of the reservoir's surface may be covered with FPV, which would lower evaporation by about 49.7% and save up to 5.9 billion cubic meters of water a year (El-Shazli & Hoermann, 2016). Similar study conducted in Brazil demonstrated that a 130 kW FPV installation reduced evaporation by 60%, highlighting the potential of FPV systems in water-stressed regions. Additionally, the cooling effect of water bodies enhances the efficiency of solar panels, that is, a 10 MW FPV system can achieve up to 10.2% higher electricity generation compared to land-based counterparts (Silva et al., 2023). Beyond energy and water benefits, FPV installations can be integrated with existing hydroelectric infrastructure to offer a complementary approach to energy generation without additional land use.

4.2.4.3 Off-Grid Solar Microgrids

Off-grid solar microgrids have emerged as a transformative solution for electrifying remote and underserved communities, particularly in Sub-Saharan Africa and South Asia. These decentralized energy systems have provided reliable electricity access to over 100 million people, fostering improvements in education, healthcare, and economic development (Baurzhan & Jenkins, 2016). For instance, in Mali, the implementation of solar microgrids has revitalized rural economies by powering businesses, schools, and healthcare facilities, despite challenges such as political instability and limited investment. The scalability and adaptability of solar microgrids make them an effective tool for achieving universal energy access, aligning with global sustainability goals (Bhattacharyya & Palit, 2021). However, to maximize their impact, supportive policies, investment in infrastructure, and community engagement are essential.

In general, the deployment of agrivoltaics, floating solar farms, and off-grid microgrids extends beyond energy generation, offering substantial social and environmental benefits. Water conservation is a notable advantage; FPV systems can reduce evaporation by up to 60%, preserving vital water resources for agriculture and consumption. Additionally, agrivoltaic systems contribute to biodiversity protection by maintaining vegetation cover, which supports pollinators and other beneficial species. Energy independence achieved through solar microgrids empowers communities, reducing reliance on fossil fuels and enhancing resilience against energy price fluctuations. Collectively, these innovations support multiple UNSDGs, including clean energy access, climate action, and sustainable land and water management.

4.3 FINANCIAL INNOVATIONS AND GOVERNANCE STRATEGIES

The global transition to sustainable energy systems is increasingly supported by innovative financial mechanisms and robust governance frameworks. The present section explores the importance of green bonds, carbon credits, and ESG-aligned investments in catalysing the growth of solar energy. Additionally, it highlights the pivotal role of PPPs in mobilizing large-scale capital and technical expertise for solar technology deployment. Critical governance challenges including transparency in ESG reporting, regulatory adaptation, and standardized impact metrics are also addressed. A thorough examination of the institutional and financial tactics influencing solar energy and global sustainability is provided in the section that follows.

4.3.1 Rise of ESG Investment

The global investment landscape has witnessed a significant shift towards ESG considerations, with solar energy emerging as a focal point for sustainable investments. In 2023, global investments in solar energy surpassed $320 billion, marking a substantial increase from previous years. This surge is largely attributed to ESG mandates from institutional investors and governmental policies aimed at promoting renewable energy sources (Obaideen et al., 2021). Institutional investors are increasingly integrating ESG criteria into their investment strategies, recognizing the long-term value and risk mitigation associated with sustainable assets. Governments worldwide have also played a pivotal role by implementing policies and incentives that encourage investment in solar energy projects. For instance, tax credits, subsidies, and renewable energy targets have created a conducive environment for both public and private investments in the solar sector (Sun et al., 2024).

The alignment of financial interests with environmental sustainability has not only accelerated the deployment of solar technologies but has also contributed to the broader goal of reducing carbon emissions. The International Energy Agency reports that the expansion of solar energy capacity has been instrumental in curbing global greenhouse gas emissions, highlighting the critical role of ESG-driven investments in combating climate change.

4.3.2 Green Bonds and Carbon Credits

Financial instruments such as green bonds and carbon credits have become essential tools in mobilizing capital for solar energy projects. Green bonds, the debt securities issued to finance environmentally friendly projects have seen exponential growth in recent years. The climate bonds initiative reports that more than $500 billion worth of green bonds were issued globally in 2023, with a sizable amount going towards solar power installations and other renewable energy projects (Yu et al., 2024b). These bonds offer investors a means to support sustainable initiatives while receiving financial returns, thereby aligning economic and environmental objectives. The proceeds from green bonds are typically earmarked for specific projects, ensuring transparency and accountability in the utilization of funds.

On the other hand, Carbon credits provide a market-based mechanism for reducing greenhouse gas emissions. By assigning a monetary value to carbon emissions, these credits incentivize companies to invest in clean energy solutions like solar power to offset their carbon footprint. The global voluntary carbon market was estimated by Ecosystem Marketplace to be worth around $2 billion in 2023, reflecting the substantial growth of the carbon credit industry. The integration of green bonds and carbon credits into the financial ecosystem has not only facilitated the scaling of solar energy projects and technologies but has also enhanced the financial viability of such initiatives (Virah-Sawmy & Sturmberg, 2025). These instruments have attracted a diverse range of investors, including pension funds, insurance companies, and impact investors, thereby broadening the capital base for solar energy development.

4.3.3 Public-Private Partnerships

PPPs have emerged as a strategic approach to accelerate the deployment of solar energy infrastructure. By leveraging the strengths of both public and private sectors, PPPs facilitate the sharing of risks, resources, and expertise, thereby enhancing the efficiency and effectiveness of solar projects. Governments often provide regulatory support, land access, and initial funding, while private entities contribute technical expertise, operational capabilities, and additional capital (Raghutla & Kolati, 2023). This collaborative model has been particularly effective in developing countries, where limited public resources necessitate private sector involvement to achieve energy access goals. For example, the World Bank's Scaling Solar programme has successfully implemented PPPs in several African countries, resulting in the rapid deployment of solar power plants at competitive tariffs. These partnerships have not only increased energy access but have also stimulated local economies by creating jobs and fostering technological innovation. Furthermore, PPPs have been instrumental in integrating solar energy into national grids, enhancing energy security, and reducing reliance on fossil fuels (Othman & Khallaf, 2023). The collaborative nature of PPPs ensures that projects are designed and implemented with a long-term perspective, aligning with both national development objectives and global sustainability goals.

4.3.4 Governance Challenges

Despite the progress made in ESG investments and the deployment of solar energy projects, several governance challenges persist, hindering the full realization of sustainable energy objectives. One of the primary challenges is ensuring transparency and consistency in ESG reporting. The lack of standardized reporting frameworks leads to discrepancies in how companies disclose their ESG performance, making it difficult for investors to assess and compare sustainability metrics. Initiatives like the Global Reporting Initiative and the Sustainability Accounting Standards Board have made strides in establishing reporting standards, but widespread adoption remains a work in progress. Regulatory support for emerging technologies, the rapid evolution of solar technologies necessitates adaptive regulatory frameworks that can accommodate innovation while ensuring safety and reliability. However, regulatory inertia and bureaucratic hurdles often delay the approval and implementation of new technologies, impeding the pace of solar energy adoption (Pasandideh & Ezaan, 2025).

Standardization in social impact measurement remains a challenge, as there is a lack of consistent metrics to evaluate the societal benefits of solar energy projects, including improvements in health, education, and economic development. This gap makes it challenging to quantify the broader benefits of solar initiatives, potentially undervaluing their contribution to sustainable development. Addressing these governance challenges requires concerted efforts from policymakers, industry stakeholders, and civil society. Establishing clear and consistent reporting standards, streamlining regulatory processes, and developing comprehensive social impact assessment tools are critical steps towards enhancing the governance of solar energy projects.

4.4 CHALLENGES AND FUTURE DIRECTIONS

4.4.1 Supply Chain Sustainability

The rapid expansion of solar PV technologies has intensified the demand for critical materials such as indium, tellurium, and gallium. These elements are integral to high-efficiency solar cells, including CdTe and CIGS technologies. However, their supply chains are fraught with challenges. For instance, indium is predominantly obtained as a by-product of zinc mining, making its availability contingent on zinc production rates. The recycling rate of indium remains alarmingly low, with studies indicating that less than 1% of indium is recycled globally, primarily due to high dissipation rates and the lack of efficient recycling technologies (Rühle, 2016). Tellurium faces similar constraints. Its primary source is as a by-product of copper refining, and its availability is subject to the fluctuations of the copper market. The International Renewable Energy Agency (IRENA) has highlighted the potential bottlenecks in the supply of such critical materials, emphasizing the need for diversified sourcing strategies and the development of alternative materials to mitigate supply risks. To address these challenges, it's imperative to invest in research focused on material substitution and to enhance recycling technologies. Developing efficient recovery methods for these scarce elements can alleviate supply pressures and contribute to the sustainability of solar PV technologies.

As the global deployment of solar panels accelerates, so does the impending challenge of managing their end-of-life. Projections by IRENA estimate that by 2050, the cumulative waste from decommissioned solar panels could reach up to 78 million tonnes. This burgeoning waste stream poses significant environmental and resource recovery challenges. Currently, the recycling infrastructure for solar panels is inadequate. A study by the NREL indicates that approximately 80% of decommissioned panels are landfilled, with only 10% being recycled and a mere 1% reused. The low recycling rates are attributed to the lack of standardized recycling protocols, economic incentives, and the technical complexities involved in recovering valuable materials like silicon, silver, and rare metals from the panels (Patel, 2025). Implementing circular economy principles is crucial to address this issue. By designing panels for easier disassembly and material recovery and by establishing robust recycling infrastructures, the industry can minimize environmental impacts and reclaim valuable resources. Moreover, policy interventions, such as Extended Producer Responsibility (EPR) schemes, can incentivize manufacturers to design recyclable products and invest in recycling technologies.

4.4.2 Equitable Financing Mechanisms

Access to solar energy remains uneven, particularly in low-income and marginalized communities. Financial barriers including high upfront costs and limited access to credit, hinder the adoption of solar technologies in these regions. Innovative financing models are essential to bridge this gap. Micro-finance institutions have emerged as pivotal players in facilitating solar adoption. Likewise, community-based financing models, such as cooperatives and collective ownership structures, empower communities to invest in and benefit from solar projects. These inclusive financing mechanisms not only democratize access to clean energy but also stimulate local economies by creating jobs and fostering community engagement. To scale these models, it's vital to provide policy support, capacity-building programmes, and risk mitigation instruments that encourage financial institutions to invest in underserved markets.

4.4.3 Policy Recommendations

To ensure the sustainable growth of the solar energy sector, comprehensive policy interventions are necessary:

- Mandate ESG disclosure for solar investments: Requiring companies to disclose ESG metrics can enhance transparency and accountability. This practice enables investors to assess the sustainability performance of solar projects, fostering responsible investment decisions.
- Subsidies for R&D in recycling technologies: Government incentives can accelerate the development of advanced recycling technologies, making the processing of end-of-life solar panels more efficient and economically viable. Such subsidies can stimulate innovation and reduce the environmental footprint of solar energy systems.
- Support for local manufacturing and job creation: Encouraging domestic production of solar components through tax incentives and infrastructure investments can bolster local economies and reduce dependency on imports. This strategy not only creates employment opportunities but also enhances the resilience of the solar supply chain.
- Implement EPR schemes: EPR policies hold manufacturers accountable for the entire life cycle of their products, including disposal and recycling. By internalizing the environmental costs, EPR can drive the design of more sustainable and recyclable solar panels.
- Facilitate access to affordable financing: Establishing PPPs and green banks can provide low-interest loans and financial guarantees for solar projects, particularly in underserved communities. Such initiatives can lower financial barriers and promote equitable access to clean energy.

By adopting these policy measures, governments can create an enabling environment that supports the sustainable expansion of solar energy, addresses environmental concerns, and ensures inclusive access to clean power.

4.5 CONCLUSION

The resolution of important ESG issues and the promotion of global sustainability objectives are being greatly aided by recent developments in solar energy technologies. Highly efficient PV systems (perovskite) and tandem cells with efficiencies surpassing 30% are redefining performance benchmarks. Concurrently, CSP systems integrated with thermal storage enhance dispatch ability and improve grid resilience. AI enabled solar forecasting and ray tracing methodologies now achieve predictive accuracies exceeding 90% as well as optimizing system-level performance and resource allocation. On the financial front, ESG-aligned instruments which include green bonds and carbon credits have mobilized over $300 billion in 2023, indicating increased investor confidence in low-carbon technologies. Moreover, PPPs are accelerating the deployment of solar infrastructure across diverse economies. However, persistent challenges such as material sustainability, recycling frameworks, and equitable financing must be addressed through coordinated policy and innovation strategies. The convergence of technological progress, capital markets, and governance frameworks position solar energy as a fundamental enabler of the global energy transition. Sustained interdisciplinary collaboration, ethical resource management, and adaptive policy mechanisms will be essential to ensuring that solar technologies remain central to achieving long-term climate objectives. The trajectory of solar innovation, therefore, is not merely technical but strategic, shaping the architecture of a resilient, inclusive, and decarbonized global economy.

REFERENCES

Afroz, M., Ratnesh, R. K., Srivastava, S., & Singh, J. (2025). Perovskite solar cells: Progress, challenges, and future avenues to clean energy. *Solar Energy*, 287(September 2024), 113205. https://doi.org/10.1016/j.solener.2024.113205

Baurzhan, S., & Jenkins, G. P. (2016). Off-grid solar PV: Is it an affordable or appropriate solution for rural electrification in Sub-Saharan African countries? *Renewable and Sustainable Energy Reviews*, 60, 1405–1418. https://doi.org/10.1016/j.rser.2016.03.016

Bhattacharyya, S. C., & Palit, D. (2021). A critical review of literature on the nexus between central grid and off-grid solutions for expanding access to electricity in Sub-Saharan Africa and South Asia. *Renewable and Sustainable Energy Reviews*, 141(April 2020), 110792. https://doi.org/10.1016/j.rser.2021.110792

Cellura, M., Luu, L. Q., Guarino, F., & Longo, S. (2025). A review on life cycle assessment of concentrating solar energy technologies. *Renewable Energy*, 249(April), 123203. https://doi.org/10.1016/j.renene.2025.123203

Corona, B., Rúa, C. de la, & San Miguel, G. (2016). Socio-economic and environmental effects of concentrated solar power in Spain: A multiregional input output analysis. *Solar Energy Materials and Solar Cells*, 156, 112–121. https://doi.org/10.1016/j.solmat.2016.03.014

Costa, D., De Regel, S., Espadas-Aldana, G., Laget, H., Kishore, R., Meuret, Y., & Duerinckx, F. (2024). Environmental and economic impacts of photovoltaic integration in concentrated solar power plants. *Solar Energy*, 274(February), 112550. https://doi.org/10.1016/j.solener.2024.112550

El-Shazli, A., & Hoermann, G. (2016). Development of storage capacity and morphology of the Aswan High Dam Reservoir. *Hydrological Sciences Journal*, 61(14), 2639–2648. https://doi.org/10.1080/02626667.2016.1151979

Garrod, A., Hussain, S. N., & Ghosh, A. (2024). The technical and economic potential for crop based agrivoltaics in the United Kingdom. *Solar Energy*, 277(June), 112744. https://doi.org/10.1016/j.solener.2024.112744

Helveston, J. P., He, G., & Davidson, M. R. (2022). Quantifying the cost savings of global solar photovoltaic supply chains. *Nature*, 612(7938), 83–87. https://doi.org/10.1038/s41586-022-05316-6

Hussain, K., & Gagliardi, A. (2022). Modelling tandem/multi-junction hybrid perovskite–organic solar cells: A combined drift–diffusion and kinetic Monte Carlo study. *Solar Energy*, 243(August), 193–202. https://doi.org/10.1016/j.solener.2022.07.033

Jignesh, P., Pujak, S., Aejaj, S., Jha, V., & Patel, J. (2018). Investigation on thermal performance of solar air heater by using artificial roughness – A review. *International Journal of Scientific Research in Science, Engineering and Technology*, 4(5), 12574–12579.

Kim, H., Cha, K., Fthenakis, V. M., Sinha, P., & Hur, T. (2014). Life cycle assessment of cadmium telluride photovoltaic (CdTe PV) systems. *Solar Energy*, 103, 78–88. https://doi.org/10.1016/j.solener.2014.02.008

Kumar, A., Prakash, O., & Dube, A. (2017). A review on progress of concentrated solar power in India. *Renewable and Sustainable Energy Reviews*, *79*(February 2016), 304–307. https://doi.org/10.1016/j.rser.2017.05.086

Ledmaoui, Y., El Maghraoui, A., El Aroussi, M., Saadane, R., Chebak, A., & Chehri, A. (2023). Forecasting solar energy production: A comparative study of machine learning algorithms. *Energy Reports*, 10, 1004–1012. https://doi.org/10.1016/j.egyr.2023.07.042

Lorenzo, E. (2021). On the historical origins of bifacial PV modelling. *Solar Energy*, 218(September 2020), 587–595. https://doi.org/10.1016/j.solener.2021.03.006

Majumdar, D., & Pasqualetti, M. J. (2018). Dual use of agricultural land: Introducing 'agrivoltaics' in Phoenix Metropolitan Statistical Area, USA. *Landscape and Urban Planning*, *170*(May 2017), 150–168. https://doi.org/10.1016/j.landurbplan.2017.10.011

Mamodiya, U., Kishor, I., Garine, R., Ganguly, P., & Naik, N. (2025). Artificial intelligence based hybrid solar energy systems with smart materials and adaptive photovoltaics for sustainable power generation. *Scientific Reports*, *15*(1), 17370. https://doi.org/10.1038/s41598-025-01788-4

Nagaraja, M. R., Biswas, W. K., & Selvan, C. P. (2025). Advancements and challenges in solar photovoltaic technologies: enhancing technical performance for sustainable clean energy – A review. *Solar Energy Advances*, *5*(September 2024), 100084. https://doi.org/10.1016/j.seja.2024.100084

NREL. (2023a). *Solar Cell Efficiency Table Guide*. 7. https://www.nlr.gov/media/docs/libraries/pv/nrel-record-cell-efficiency-data-table-guide.pdf

NREL. (2023b). *Highest Confirmed Conversion Efficiencies for Research Cells for a Range of Photovoltaic Technologies*. https://www.nlr.gov/pv/cell-efficiency

Obaideen, K., Nooman AlMallahi, M., Alami, A. H., Ramadan, M., Abdelkareem, M. A., Shehata, N., & Olabi, A. G. (2021). On the contribution of solar energy to sustainable developments goals: Case study on Mohammed bin Rashid Al Maktoum Solar Park. *International Journal of Thermofluids*, 12, 100123. https://doi.org/10.1016/j.ijft.2021.100123

Othman, K., & Khallaf, R. (2023). Renewable energy public-private partnership projects in Egypt: Perception of the barriers and key success factors by sector. *Alexandria Engineering Journal*, 75, 513–530. https://doi.org/10.1016/j.aej.2023.06.009

Pasandideh, A. S., & Ezaan, G. (2025). Governance challenges of renewable energy development in fossil fuel dependent countries with a focus on Iran. *Energy for Sustainable Development*, 87, 101759. https://doi.org/https://doi.org/10.1016/j.esd.2025.101759

Patel, J. (2025). *Studies on Operation of Large Scale Solar Flat Plate Reflecting System and Its Thermal Performance*. https://doi.org/10.13140/RG.2.2.14423.38560

Patel, J., & Patel, A. R. (2023). Effect of octagonal shaped aperture area on optical performance of the simple flat plate solar reflecting system. *2023 IEEE Renewable Energy and Sustainable E-Mobility Conference (RESEM)*, 1–6. https://doi.org/10.1109/RESEM57584.2023.10236285

Patel, J., & Patel, A. R. (2023). Effect of reflector height to base ratio on the optical performance of square shaped aperture area based flat plat solar reflecting system. *Proceeding of International Conference on Recent Advances in (Applied) Sciences & Engineering (Raise) : Vol.* IV. ISBN: 978–81–962938-1–9, 9–16.

Patel, J., & Patel, A. R. (2024a). Numerical investigation on the effect of different types of configuration on the optical performance of solar flat plat reflecting system. *Sustainable Energy Technologies and Assessments*, 61, 103586. https://doi.org/https://doi.org/10.1016/j.seta.2023.103586

Patel, J., & Patel, A. R. (2024b). Recent developments in large-scale solar flat plate reflecting systems: Optical analysis using specialized numerical and analytical tools. In *The Costs of Climate Change Mitigation Innovations* (eds.) David S-K. T., Jacqueline A. S. (pp.173–212). CRC Press.

Patel, J., Patel, A. R., & Mulasiya, C. (2023). Determination of optical and thermal performance of a large - sized flat plate solar reflector unit under azimuthal sun alignment. *Journal of Thermal Analysis and Calorimetry*, *148*, 8991–9011. https://doi.org/10.1007/s10973-023-12282-2

Patel, J., Patel, A. R., & Tyagi, H. (2024). Advances in the solar thermal systems assisted production of green hydrogen: Its analysis, scaling-up techniques, and economics aspects as applied to tropical regions. In R. K. S. Singh, Paramvir, Avinash Kumar Agarwal, Anupma Thakur (Ed.), *Challenges and Opportunities in Green Hydrogen Production* (pp. 239–267). Springer Nature. https://doi.org/10.1007/978-981-97-1339-4_10

Patel, J., & Singh, V. (2015a). *Energy Storage System for Cooling and Heating Application.* Thesis for Master of Engineering, A.D. Patel Institute of Technology, Gujarat, India.

Patel, J., & Singh, V. (2015b). PCM augumented composite wall for base transceiver station (BTS) for telecom towers. *Proceedings of the 17th ISME Conference ISME17 October 3–4, 2015, IIT* Delhi, New Delhi, 3–8.

Priyanshu, S., Abhishek, S., Akshay, T., Vijay, T., & Patel, J. (2018). Development of multipurpose commercial solar furnace - A review. *International Journal of Scientific Research in Science, Engineering and Technology*, 4, 579–585.

Raghutla, C., & Kolati, Y. (2023). Public-private partnerships investment in energy as new determinant of renewable energy: The role of political cooperation in China and India. *Energy Reports*, 10(October), 3092–3101. https://doi.org/10.1016/j.egyr.2023.09.139

Rühle, S. (2016). Tabulated values of the Shockley-Queisser limit for single junction solar cells. *Solar Energy*, 130, 139–147. https://doi.org/10.1016/j.solener.2016.02.015

Sahu, P. K., Roy, J. N., & Chakraborty, C. (2023). Performance assessment of a bifacial PV system using a new energy estimation model. *Solar Energy*, 262(June), 111818. https://doi.org/10.1016/j.solener.2023.111818

Silva, D. D., Cardoso, E. M., Basquerotto, C., Pereira, J. A., Turra, A. E., & Feldhaus, J. (2023). Outlook on the Brazilian scenario of floating photovoltaic solar energy. *Energy Reports*, 10(October), 4429–4435. https://doi.org/10.1016/j.egyr.2023.11.004

Steffen, B. (2018). The importance of project finance for renewable energy projects. *Energy Economics*, 69, 280–294. https://doi.org/10.1016/j.eneco.2017.11.006

Sun, Y., Rahman, M. M., Xinyan, X., Siddik, A. B., & Islam, M. E. (2024). Unlocking environmental, social, and governance (ESG) performance through energy efficiency and green tax: SEM-ANN approach. *Energy Strategy Reviews*, 53(March), 101408. https://doi.org/10.1016/j.esr.2024.101408

Supe, H., Abhishek, A., & Avtar, R. (2024). Assessment of the solar energy–agriculture–water nexus in the expanding solar energy industry of India: An initiative for sustainable resource management. *Heliyon*, 10(1), e23125. https://doi.org/10.1016/j.heliyon.2023.e23125

Venkatesh, R., Singh, P. K., Bashir, M. N., Lee, J. S., Yaswanth, K. K., Soudagar, M. E. M., Hossain, I., Obaid, S. Al, & Alharbi, S. A. (2024). Performance study of cadmium telluride solar cell featured with silicon thin film made by sol-gel route. *Silicon, 17*, 191–203. https://doi.org/10.1007/s12633-024-03179-2

Venkateswaran, D., & Cho, Y. (2024). Efficient solar power generation forecasting for greenhouses: A hybrid deep learning approach. *Alexandria Engineering Journal*, 91(February), 222–236. https://doi.org/10.1016/j.aej.2024.02.004

Virah-Sawmy, D., & Sturmberg, B. (2025). Socio-economic and environmental impacts of renewable energy deployments: A review. *Renewable and Sustainable Energy Reviews, 207*(September 2024), 114956. https://doi.org/10.1016/j.rser.2024.114956

Yao, Y., Lv, F., Zhao, X., Yang, X., Wu, B., Li, P., & Song, Q. (2023). A monolithic all-perovskite tandem solar cell with 2-T, 3-T and 4-T architecture integrated. *Solar Energy Materials and Solar Cells*, 259(July), 2–7. https://doi.org/10.1016/j.solmat.2023.112376

Yu, Q., Yongfeng, M., Jiaheng, H., & Yunci, Y. (2024a). Solar space thermal energy utilization and AI navigation based on light sensors in museum visual communication design. *Thermal Science and Engineering Progress*, 54, 102874. https://doi.org/https://doi.org/10.1016/j.tsep.2024.102874

Yu, Z., Farooq, U., Alam, M. M., & Dai, J. (2024b). How does environmental, social, and governance (ESG) performance determine investment mix? New empirical evidence from BRICS. *Borsa Istanbul Review*, 24(3), 520–529. https://doi.org/10.1016/j.bir.2024.02.007

Vadhera, J., Sura, A., Nandan, G., Dwivedi, G., (2018). Study of phase change materials and its domestic application. *Materialstoday: Proceedings, 5*(2) Part 1, 3411–3417.

5 Internet of Things in Enabling Waste-to-Energy for Fuel Cell Technology in Industrial Settings

Johnson Kehinde Abifarin, David Oyeyemi Aboyeji, Fredah Batale Abifarin, and Joseph Chukwuka Ofodu

5.1 INTRODUCTION

5.1.1 Overview of Global Energy Demand and Industrial Waste Generation

Due to factors such as urbanisation, industrialisation, and population growth, the world's energy demand has been steadily increasing. Global energy consumption rose from 537.5 quadrillion British thermal units (Btu) in 2010 to 620.0 quadrillion Btu in 2018, according to the International Energy Agency (IEA). At a 1.2% annual growth rate, projections indicate that global energy consumption will reach 910.7 quadrillion Btu by 2050. This increase is especially noticeable in developing nations, where rising living standards and fast industrialisation greatly increase the demand for energy [1, 2].

At the same time, industrial operations play a significant role in the production of waste worldwide. Mini businesses and households often generate significant waste, such as animal wastes, fruit wastes, Agro wastes, etc., and efforts have been geared towards converting these wastes into high-value products [3–5]. Over half of the 9.2 billion tonnes of industrial waste generated in 2011 came from Asia. On average, 1.74 tonnes of industrial waste are generated per person each year. Despite regional and sector-specific variations in the volume and composition of industrial waste, the trend shows a steady rise in tandem with industrial growth [6].

The Internet of Things (IoT) plays a critical role in enhancing waste-to-energy (WtE) systems by enabling real-time monitoring, data-driven decision-making, and automated control across the entire energy conversion process [7]. In this context, IoT facilitates smarter waste classification, optimises fuel cell performance, ensures efficient syngas conditioning, and supports predictive maintenance. By integrating IoT, WtE systems become more responsive, energy efficient, and adaptable, aligning with Industry 4.0 and sustainability goals [8, 9].

DOI: 10.1201/9781003671039-5

There are growing opportunities and challenges at the intersection of rising industrial waste generation and increasing energy demand. WtE technologies, enhanced by fuel cell integration and IoT-enabled smart systems, offer innovative pathways to convert waste into clean, usable energy. These systems not only support energy diversification and low-carbon industrial practices but also enable real-time optimisation, monitoring, and control across the WtE value chain [10]. As industries move towards more sustainable and intelligent infrastructure, the fusion of IoT and fuel cell-based WtE solutions stands out as a key enabler of circular economy goals and long-term energy resilience.

5.1.2 Importance of Sustainable Waste Management and Clean Energy

Numerous works have been done on transforming wastes into valuable products, such as biomedical products, aerospace, lubricants, energy, automotive, etc. [11–17]. Clean energy and sustainable waste management are essential for tackling the growing environmental problems brought on by urbanisation and industrialisation [18–20]. Reducing waste accumulation and producing renewable energy are two advantages of integrating WtE technologies. According to recent research, sophisticated WtE systems, like plasma gasification, can attain energy efficiencies of up to 60%, highlighting their potential for producing sustainable energy [21].

Additionally, the transition to a circular economy underscores the importance of reducing, reusing, and recycling waste to conserve resources and minimise environmental impact. By utilising waste as a valuable energy resource, sustainable waste management techniques not only solve the urgent problem of waste disposal but also promote energy security. This all-encompassing strategy encourages economic resilience and environmental stewardship, which is in line with global sustainability goals [22].

5.1.3 Role of Fuel Cell Technology in Waste-to-Energy Systems

Fuel cell technology is essential to WtE systems because it effectively transforms organic waste materials into energy forms that can be used, like hydrogen and electricity, while also solving waste management issues [23].

Microbial fuel cells (MFCs), one of the different fuel cell technologies, have attracted a lot of interest due to their ability to treat wastewater and produce electricity. MFCs work by using electrochemically active microorganisms to oxidise organic substrates found in waste streams. This process creates electrons, which then flow through an external circuit to produce electricity. In addition to lowering wastewater's chemical oxygen demand (COD), this process provides a sustainable way to recover energy [24, 25].

MFCs are now more applicable in a wider range of industrial settings thanks to improved configurations brought about by advancements in MFC design, such as single- and double-chamber systems with air cathodes. These developments have increased MFCs' capacity to treat a wider range of wastewater types, including those from residential, commercial, industrial, and municipal sources [26, 27].

Within WtE systems, microbial electrolysis cells (MECs) are an additional aspect of fuel cell technology in addition to MFCs. MECs provide a promising path towards the production of biohydrogen by electrochemically converting organic matter into hydrogen gas through the application of an external voltage. This hydrogen can then be used in internal combustion engines or proton exchange membrane (PEM) fuel cells, which will help create a cleaner energy environment [28].

A sustainable method of waste management and energy production is offered by the incorporation of fuel cell technologies into WtE systems. These systems, which effectively transform organic waste into valuable energy resources by utilising the biochemical processes of microorganisms, are in line with international initiatives to lessen environmental pollution and support renewable energy sources [29, 30].

5.1.4 Objectives and Scope

The goal of this review is to give a thorough overview of the global energy scene and the expanding problem of industrial waste generation. It emphasises how crucial it is to manage waste sustainably and move towards clean energy sources for industrial applications. Figure 5.1 gives a general overview of the importance of WtE technologies. Examining the fundamentals, varieties, and working mechanisms of fuel cell technologies appropriate for WtE applications is a major goal of the review. By highlighting the technical, financial, and environmental advantages of fuel cell systems, the review investigates how they can be incorporated into WtE frameworks. Additionally, it examines recent developments, case studies, and technological trends

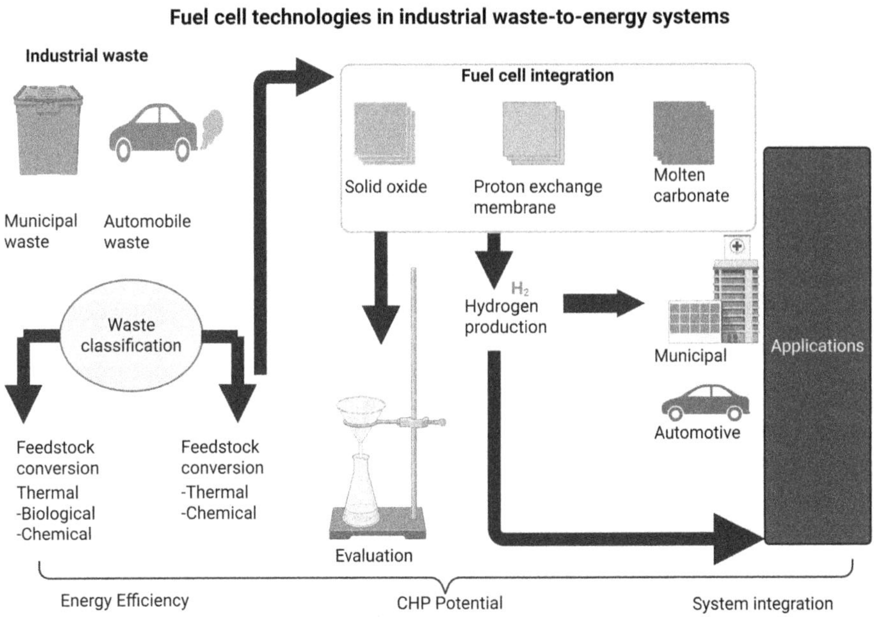

FIGURE 5.1 Fuel cell technologies in industrial WtE systems, using hydrogen as a case study.

that show how fuel cells can be effectively used to recover energy from industrial waste streams. Along with outlining potential fixes and future development directions, the review also identifies the main obstacles and constraints preventing the broad industrial adoption of fuel cell-based WtE systems.

The review's focus is on how fuel cell technologies in industrial settings relate to waste management and the generation of clean energy. The need for sustainable energy interventions is introduced by talking about the rising rate of industrial waste generation and the world's energy demand. The story then moves into a thorough examination of clean energy technologies, emphasising the ability of fuel cells like MFCs, proton exchange membrane fuel cells (PEMFCs), and MECs to turn waste into hydrogen or electricity.

This chapter presents two key innovations driving the advancement of industrial WtE systems. First, it explores the integration of fuel cell technologies, such as solid oxide, PEM, and molten carbonate fuel cells (MCFCs) into WtE processes, emphasising their design flexibility, operational efficiency, and adaptability to different industrial waste streams. This innovation enhances the potential for clean energy recovery and supports low-carbon industrial practices. Second, the chapter highlights the role of the IoT in enabling intelligent, real-time monitoring and optimisation across the WtE value chain. By incorporating IoT, these systems gain capabilities in predictive maintenance, dynamic feedstock management, and performance optimisation. Real-world case studies and recent technological advancements are analysed to demonstrate practical outcomes and effectiveness. This chapter also addresses critical issues such as policy, scale-up, and system integration, offering a forward-looking perspective on how these two innovations together can transform industrial energy systems towards greater sustainability and circularity.

5.2 WASTE STREAMS IN INDUSTRIAL SETTINGS

5.2.1 Classification of Industrial Wastes

A wide range of materials produced by manufacturing and other industrial processes is included in industrial waste. To implement efficient management and treatment plans, these wastes are usually categorised according to their physical condition and chemical makeup. Solid, liquid, gaseous, and thermal wastes are the main categories [31, 32].

Materials like packaging materials, plastics, metal scraps, and sludge from industrial processes fall under the category of solid waste. To reduce its negative effects on the environment, solid industrial waste frequently needs to be disposed of or recycled properly [33, 34].

Liquid waste is produced by a variety of industrial processes. Liquid waste is made up of effluents that contain heavy metals, organic and inorganic materials, and other impurities. To stop water pollution and save aquatic ecosystems, effective treatment is essential [35].

During combustion and chemical reactions, industries release gaseous pollutants such as sulphur dioxide, nitrogen oxides, carbon monoxide, and volatile organic compounds [36, 37]. To reduce the risks to human health and the environment, these emissions must be controlled because they contribute to air pollution. This is referred to as gaseous wastes [36, 38].

Thermal waste occurs when excess heat produced by industrial processes frequently results in thermal pollution if left unchecked. Thermal emissions can be decreased, and energy efficiency can be increased by putting waste heat recovery systems into place [39, 40].

To ensure regulatory compliance, promote sustainable industrial operations, and develop targeted waste management practices, it is essential to comprehend how industrial wastes are classified.

5.2.2 Waste Generation Patterns in Major Industries

Due to variations in raw materials, production methods, and final products, waste generation patterns differ greatly amongst the main industrial sectors. Designing efficient waste management and energy recovery strategies requires an understanding of the unique waste types and quantities produced by each industry.

Hazardous waste, such as solvents, heavy metals, and acidic or alkaline residues, is primarily produced by the chemical industry [41]. These wastes come from the processes of formulation, purification, and synthesis. Many chemical byproducts are toxic and complex. So it's crucial to treat and dispose of them properly to reduce negative effects on the environment and human health.

Flue gases, sulphur compounds, spent catalysts, and oily sludges are amongst the waste products produced in the petrochemical sector. These come from processes like reforming, cracking, and distilling crude oil. Significant heat and gaseous emissions are another consequence of this sector's high energy intensity [42, 43].

Large amounts of organic waste, including fruit peels, vegetable residues, fats, oils, and wastewater containing biodegradable materials, are frequently produced by the food processing industry. Anaerobic digestion, a biological treatment technique that can generate biogas for energy, is frequently appropriate for these wastes [44–46].

Wood chips, bark, wastewater treatment sludge, and chemical residues from pulping and bleaching are all considered waste in the paper and pulp industry. Additionally, the industry generates a substantial amount of liquid and thermal waste [47, 48]. Energy recovery from these byproducts is frequently accomplished through techniques like biomass combustion and black liquor recovery.

The selection of suitable WtE technologies, especially those incorporating fuel cell systems for clean and effective energy recovery, is aided by an understanding of the distinct waste generation patterns in these industries [49, 50].

5.2.3 Environmental and Regulatory Considerations

When implementing fuel cell technologies for WtE applications in industrial settings, environmental and regulatory factors are crucial. These factors include the advantages of fuel cells for the environment, adherence to environmental laws, and the incorporation of sustainable practices into business operations.

By treating industrial waste and producing renewable energy, fuel cells, especially MFCs, offer a twofold benefit. MFCs lessen the environmental impact of disposing of industrial waste by using the metabolic processes of microorganisms to turn organic waste into electricity. This technology meets the needs for energy recovery and waste treatment at the same time, which is consistent with the ideas of sustainability and the circular economy [51, 52].

Adoption of fuel cell technologies must, from a regulatory perspective, comply with environmental laws and standards that control emissions and waste management. Regulations that restrict the discharge of pollutants and require that waste be treated before being disposed of must be followed by industries. By offering an effective way to treat waste and lowering harmful emissions, fuel cell systems can help meet these regulatory requirements [53–55].

Furthermore, incorporating fuel cell technologies into industrial operations helps meet environmental goals established by authorities. Industries can lessen their dependency on fossil fuels, cut greenhouse gas emissions, and help mitigate climate change by turning waste into energy. This improves the environmental sustainability of industrial operations in addition to guaranteeing regulatory compliance.

In conclusion, the effective integration of fuel cell technologies into industrial WtE systems depends on regulatory and environmental factors. In accordance with environmental regulations and supporting the more general objectives of environmental preservation and sustainable development, these technologies provide a sustainable solution for energy production and waste management.

5.3 OVERVIEW OF FUEL CELL TECHNOLOGIES

5.3.1 Working Principles of Fuel Cells

Fuel cells are electrochemical devices that use redox reactions to directly transform the chemical energy of fuels into electrical energy, providing a sustainable and effective substitute for traditional combustion-based power generation [56, 57].

An anode, a cathode, and an electrolyte are the three primary parts of a fuel cell. The following actions are part of the general operation:

1. Fuel Introduction: At the fuel cell's anode side, a fuel—usually hydrogen—is introduced.
2. Oxidation Reaction at the Anode: Protons (H^+) and electrons (e^-) are produced by the catalytic splitting of hydrogen molecules at the anode. This response can be shown as:

$$H_2 \rightarrow 2H^+ + 2e^-$$

3. Proton Migration Through Electrolyte: Protons move in the direction of the cathode through the electrolyte. Charge separation is ensured by the electrolyte's design, which permits only particular ions (such as H^+) to flow through.
4. Unable to move through the electrolyte, the electrons move from the anode to the cathode via an external circuit, producing an electric current that can be used to power external tasks.

5. Reduction Reaction at the Cathode: Water is created at the cathode when incoming electrons and protons mix with oxygen molecules from the air:

$$\frac{1}{2}O_2 + 2H^+ + 2e^- \rightarrow H_2O$$

In a hydrogen fuel cell, the general reaction is [58, 59]:

$$H_2 + \frac{1}{2}O_2 \rightarrow H_2O + \text{electricity} + \text{heat}$$

Fuel cells are an environmentally friendly energy source because they produce electricity continuously with only heat and water as byproducts.

Fuel cells are appealing for contemporary energy systems because of their many benefits. They frequently outperform conventional combustion engines in terms of efficiency, especially when their waste heat is captured and used in combined heat and power (CHP) systems. Fuel cells are unique in the environmental arena because they only produce heat and water as byproducts, which leads to a considerable decrease in greenhouse gas emissions and very little pollution [60]. Furthermore, if fuel and oxidant supplies are maintained, fuel cells can run continuously, unlike batteries, guaranteeing a steady and continuous power output.

Fuel cells are being used more and more in industrial settings for a variety of energy applications, especially WtE systems. They are perfect for several applications due to their ability to effectively transform chemical energy directly into electrical energy [60]. They are used, for example, in backup power systems, which supply dependable electricity in the event of an outage. Additionally, they are employed in CHP systems, which produce both electricity and beneficial thermal energy for industrial operations. Additionally, by turning industrial waste gases into electricity, fuel cells are essential for WtE applications, which enable businesses to produce clean energy and reduce waste at the same time.

5.3.2 Types of Fuel Cells Suitable for Waste-to-Energy

A clean and effective substitute for traditional combustion-based power generation, fuel cells are electrochemical devices that directly convert the chemical energy of fuels into electrical energy through redox reactions [56, 57]. Different types of fuel cells, each with distinct features that align with operational requirements, are especially well-suited for WtE applications in industrial settings (Table 5.1).

Solid polymer electrolyte is used in PEMFCs, which function at relatively low temperatures (up to 60°C). They are perfect for applications needing variable power output and fast response because of their high power densities and short start-up times. They do, however, require pure hydrogen fuel, which may be a drawback when working with gases derived from impure waste [61, 62].

Solid oxide fuel cells (SOFCs) use a solid ceramic electrolyte and operate at temperatures between 650°C and 1000°C (Table 5.1). They are very effective for WtE applications because of their high operating temperature, which enables internal reforming of a range of fuels, including waste-derived syngas and biogas. Furthermore, cogeneration can make use of the high-grade waste ssheat generated [56, 63].

TABLE 5.1
Fuel cells suitable for WtE applications

Fuel Cell Type	Electrolyte	Fuel Examples	Operating Temperature	Electrical Efficiency (LHV)	Applications	Advantages	Challenges	Reference
Polymer Electrolyte Membrane (PEM)	Perfluorosulfonic acid	Hydrogen (H_2) (pure or from reforming methane, methanol); low tolerance to CO and impurities	<120°C	40–60%	Transportation, Portable Power, Backup Power	Quick start-up, Low operating temperature	Expensive catalysts, sensitive to fuel impurities	[61, 62]
Alkaline Fuel Cell (AFC)	Aqueous potassium hydroxide	Hydrogen (H_2) (very pure); ammonia (cracked); sensitive to CO_2 contamination	90–100°C	Nearly 70%	Military, Space Applications	High efficiency, Quick start-up	CO_2 sensitivity, electrolyte management	[65, 66]
Phosphoric Acid Fuel Cell (PAFC)	Phosphoric acid	Hydrogen (H_2) (can tolerate small amounts of CO); reformed natural gas or biogas	130–180°C	~46%	Distributed Generation, CHP	Tolerance to fuel impurities	Long start-up time, sulphur sensitivity	[75]
Molten Carbonate Fuel Cell (MCFC)	Molten carbonates	Hydrogen (H_2); carbon monoxide (CO); methane (CH_4); biogas, syngas	600–700°C	50–60%	Electric Utility, Distributed Generation	High efficiency, Fuel flexibility	High-temperature corrosion, long start-up time	[64]
Solid Oxide Fuel Cell (SOFC)	Yttria-stabilised zirconia	Hydrogen (H_2); carbon monoxide (CO); methane (CH_4); syngas, ammonia, light hydrocarbons	650–1,000°C	Up to 60%	Auxiliary Power, Electric Utility	High efficiency, Fuel flexibility	High-temperature operation, material degradation	[56, 63]

Molten carbonate salt is used as the electrolyte in MCFCs, which likewise function at high temperatures (600–700°C), as shown in Table 5.1. Because they can directly use carbon-rich fuels like landfill gas and capture carbon dioxide emissions, they are especially well-suited for large-scale industrial WtE applications, which lower greenhouse gas outputs [64].

Alkaline Fuel Cells (AFCs) use an aqueous solution of potassium hydroxide as the electrolyte and function at lower temperatures (roughly 90–100°C) [65, 66]. They have a high electrical efficiency and have been utilised in space applications in the past. Their applicability in WtE scenarios, where fuel purity cannot be guaranteed, is limited by their sensitivity to carbon dioxide, which requires the use of pure hydrogen and oxygen [67, 68].

In short, SOFCs and MCFCs are better suited for industrial applications because they can generally adapt better to the various and impure fuel streams that are typical of WtE processes. Despite their efficiency, PEMFCs and AFCs need purer fuel sources, which may call for extra processing steps when utilised in WtE systems.

5.3.3 Key Performance Parameters

When assessing the suitability of different fuel cell technologies for WtE applications, especially in industrial settings, key performance parameters are crucial. These factors, which together affect a fuel cell system's operational viability, energy yield, and environmental impact, include efficiency, operating temperature, and fuel flexibility.

One of the most important fuel cell performance indicators is efficiency. Fuel cells achieve higher electrical efficiencies than conventional combustion systems because they use electrochemical reactions to directly convert chemical energy into electrical energy. For example, when incorporated into CHP systems, SOFCs and MCFCs can achieve electrical efficiencies of 45–60%, which may surpass 80%. This is significantly greater than the typical efficiency range of conventional thermal power plants, which is between 30% and 40% [69].

Fuel cell application domain and material selection are influenced by operating temperature. The high operating temperatures of SOFCs and MCFCs (500–1000°C) facilitate the internal reforming of fuels based on hydrocarbons, such as syngas produced from industrial waste. AFCs and PEMFCs function at lower temperatures (60–100°C), which makes them more appropriate for portable and backup power applications but less adaptable to impure or mixed fuel streams [70].

The ability of a fuel cell to run on a range of fuels is known as fuel flexibility. Given that industrial waste frequently produces heterogeneous fuel gases, such as syngas, biogas, and other organic compounds, this feature is crucial for WtE applications. High-temperature fuel cells, such as SOFCs and MCFCs, can process fuels with a high tolerance to impurities like carbon monoxide and methane and exhibit significant fuel flexibility. However, PEMFCs and AFCs need more refined hydrogen sources and are typically less resilient to pollutants like sulphur compounds and CO_2.

In short, the viability and scalability of fuel cell systems in WtE applications are significantly influenced by the three main performance parameters: efficiency, operating temperature, and fuel flexibility. Because of their greater fuel tolerance and ability to integrate with current heat recovery systems, high-temperature fuel cells are typically better suited for industrial WtE.

5.3.4 Comparative Analysis of Fuel Cell Types in Industrial Contexts

Several factors, including fuel type, operating conditions, efficiency, and scalability, influence the choice of fuel cell technology in industrial WtE applications. As shown in Table 5.1, each of the following fuel cell types—PEMFCs, SOFCs, MCFCs, and AFCs—offers special characteristics that influence how well they convert industrial waste into useful energy.

PEMFCs are renowned for their quick start-up times and low operating temperatures, which make them perfect for mobile or decentralised industrial power systems. Their high sensitivity to fuel impurities and usual need for purified hydrogen limit their direct use in most WtE setups, where waste-derived fuels may contain contaminants. Conversely, SOFCs can use a wider variety of fuels, such as syngas, biogas, and other carbon-based waste gases, because they operate effectively at high temperatures, frequently above 700°C. Because of this, SOFCs are especially well-suited for stationary industrial operations that require CHP capabilities and fuel flexibility [71, 72].

In addition to operating at high temperatures, MCFCs are unique in that they can use fuels high in carbon while also absorbing carbon dioxide during internal operations. They are appealing for heavy industrial settings looking to generate power and reduce emissions due to their large-scale potential and capacity to operate with raw fuel streams like landfill gas. Despite their high electrical efficiency, AFCs are less common in industrial settings because of their sensitivity to carbon dioxide and requirement for extremely pure input gases. Their use is mostly restricted to specialised or regulated environments, like aerospace applications [73, 74].

Because of their ability to withstand high energy demands and their tolerance for fuel impurities, SOFCs and MCFCs provide more flexibility and resilience in industrial WtE scenarios. Although PEMFCs and AFCs are still useful in some situations, their direct integration into many waste-intensive industrial processes is limited by their reliance on cleaner fuels.

5.4 CONVERSION OF WASTE TO FUEL CELL FEEDSTOCK

5.4.1 Thermal, Biological, and Chemical Conversion Processes

Waste is converted into fuel cell feedstock through several chemical, biological, and thermal processes that produce hydrogen-rich gases that can be used in fuel cell applications. These techniques are essential to sustainable energy plans, particularly in industrial environments where significant waste production occurs [76, 77].

5.4.1.1 Thermal Conversion Processes

Syngas, a mixture mostly made up of hydrogen and carbon monoxide, is created when organic materials are broken down at high temperatures using thermal processes like gasification and pyrolysis [78]. Gasification works well for processing industrial byproducts, agricultural residues, and municipal solid waste because it operates in an oxygen-limited environment. Syngas, bio-oil, and biochar are the products of pyrolysis, which is carried out without oxygen [79, 80]. Hydrogen for fuel cells can be extracted from these products through additional refinement. By combining these thermal processes with fuel cell systems, the environmental impact is decreased, and energy recovery efficiency is increased.

5.4.1.2 Biological Conversion Processes

In the biological process known as anaerobic digestion, food scraps and sewage sludge are broken down by microorganisms without oxygen to create biogas, which is a combination of carbon dioxide and methane [81]. Fuel cells made to handle methane-rich gases can use this biogas directly, or it can be purified to extract hydrogen. In addition to producing renewable energy, the process lowers the amount of waste that needs to be disposed [82, 83].

5.4.1.3 Chemical Conversion Processes

Hydrocarbons are transformed into hydrogen-rich gases by chemical processes, especially reforming techniques. One common technique is steam methane reforming (SMR), in which hydrogen is created when methane and steam react at high temperatures [84]. Anaerobic digestion biogas can also be reformatted to produce fuel cell-compatible hydrogen. These chemical reactions are essential for converting different waste-derived gases into high-purity hydrogen [83, 85].

5.4.2 Hydrogen Production from Waste Streams

A crucial route in sustainable energy systems is the production of hydrogen from waste streams, particularly in the context of fuel cell-based WtE applications. Waste materials can be converted into hydrogen-rich gases that can be used in fuel cells using a variety of conversion technologies, such as thermal, biological, and chemical processes. These procedures greatly reduce waste and protect the environment in addition to providing a way to recover energy [76, 86].

Syngas, a mixture of hydrogen and carbon monoxide, is produced by thermal conversion techniques like gasification and pyrolysis, which break down organic and industrial waste at high temperatures. Plastics, agricultural residues, and municipal solid waste have all been successfully processed using gasification, which takes place in carefully regulated oxygen-limited environments. Without oxygen, pyrolysis produces hydrogen-containing syngas, bio-oil, and char that can be refined or treated further. Studies have demonstrated that combining these thermal systems with fuel cell technologies improves energy recovery and lowers emissions, making the process environmentally friendly and energy efficient [87, 88].

Microbial activity breaks down organic waste products in an oxygen-free environment as part of biological conversion, especially anaerobic digestion. Biogas is produced by this process and is mostly made up of carbon dioxide and methane. Biogas can be transformed into hydrogen by partial oxidation or reforming. Numerous studies have shown that this pathway not only generates renewable hydrogen but also enhances the results of waste treatment, particularly for food waste and sludge produced by the beverage and food industries [89, 90].

Hydrocarbons in biogas or syngas are frequently converted to hydrogen using chemical reforming techniques like steam reforming and autothermal reforming. Because of its high hydrogen yield, SMR is still the most widely used technique. Recent studies have extended its use to biogas and other waste-derived gases. These techniques produce high-purity hydrogen, which is essential for effective fuel cell operation, but they usually call for catalysts and high-temperature conditions [23, 91].

Overall, incorporating these waste conversion pathways into systems for producing hydrogen is a significant step towards attaining clean energy generation and sustainable waste management in industrial settings. It is a prime example of a circular economy, in which waste is turned into a useful resource as we move towards low-carbon energy.

5.4.3 Syngas Cleaning and Conditioning for Fuel Cell Applications

Syngas needs to be thoroughly cleaned and conditioned before it can be used effectively in fuel cell systems, particularly in delicate technologies like SOFCs or PEMFCs. A range of contaminants can impair fuel cell performance or contaminate the catalysts in raw syngas, which is usually created by thermal conversion techniques like gasification or pyrolysis of biomass and industrial waste. Particulates, tars, sulphur compounds (like hydrogen sulphide), halides, nitrogen compounds (like ammonia and HCN), and trace metals are some examples of these impurities [92, 93].

The elimination of tar compounds and solid particles is the first step in the syngas cleaning process. Scrubbers, filters, and cyclones are frequently employed to get rid of these pollutants. Because they can condense and clog downstream equipment, tars are especially problematic. To improve fuel quality and energy yield, advanced tar reforming techniques—such as catalytic cracking or thermal reforming at high temperatures—are being used more and more to transform tars into additional syngas components [94, 95].

Sulphur compounds are usually the next target after particulate and tar removal. Since platinum-based catalysts in PEMFCs are extremely vulnerable to sulphur poisoning, these are harmful to most fuel cell catalysts. Adsorbents that can capture sulphur species even at high temperatures, such as activated carbon or zinc oxide, are frequently used to achieve desulphurisation. Although SOFC systems have somewhat higher sulphur tolerances, prolonged exposure still calls for cleaning to increase efficiency and operational life [96].

To maintain optimal performance in various fuel cell types, gas conditioning procedures may also involve adjusting the hydrogen-to-carbon monoxide (H_2/CO) ratio [23, 97]. For this, water-gas shift reactors are frequently used, particularly when a higher hydrogen concentration is needed. Furthermore, because trace elements like heavy metals and chlorides can corrode system components or poison catalysts, their removal is crucial.

In the end, the durability, effectiveness, and viability of incorporating fuel cell technologies into WtE systems are greatly influenced by the quality of syngas. This process can now be streamlined, thanks to recent advancements in integrated gas cleaning modules, which increase the viability of syngas as a feedstock for clean energy production in industrial settings.

5.4.4 Integration Challenges and Pretreatment Needs

The complex and variable nature of waste materials and the strict fuel quality requirements of fuel cells make it difficult to integrate waste-derived feedstocks into fuel cell systems. The inconsistent waste composition is one of the main issues since it has a direct impact on the yield and make-up of the resulting syngas or biogas. To guarantee a steady and clean fuel supply that satisfies the operational requirements of the target fuel cell technology, such variability calls for strong and flexible pretreatment systems.

To eliminate moisture, level out particle size, and lessen the number of inert materials or possible pollutants, pretreatment procedures are crucial. To guarantee effective conversion and avoid excessive tar formation, feedstocks in thermal processes such as gasification or pyrolysis, for instance, frequently need to be dried to lower their moisture content below 20%. Furthermore, to maintain consistent heating and reaction rates during conversion, physical pretreatment techniques like shredding or milling are employed to standardise particle size [98, 99].

Depending on the feedstock, chemical and biological pretreatment might also be required. For example, complex organics can be broken down and biogas yields increased in anaerobic digestion processes by using pretreatment techniques like thermal hydrolysis or enzymatic treatments. Like this, chemical leaching can be used to eliminate heavy metals or chlorine from waste streams, especially those containing electronic or municipal solid waste, which are increasingly being explored as possible sources of hydrogen-rich fuels [100].

Matching the input requirements of the fuel cell type with the output fuel characteristics is a crucial integration challenge. SOFCs are more resilient but still need gas cleaning to prevent long-term deterioration, while PEMFCs need high-purity hydrogen with little CO, sulphur, or other impurities. The pretreatment procedure must therefore be planned in accordance with the requirements of the fuel cell system, which frequently calls for a modular or hybrid approach.

The integrated system's economic viability and scalability present additional challenges. Numerous pretreatment and cleaning procedures raise the initial and ongoing expenses, making them unaffordable for small- and medium-sized industrial facilities. Furthermore, for integration to work, the waste conversion unit and the fuel cell must be compatible in terms of physical interface and control systems that can manage changes in fuel quality, pressure, and dynamic loads.

A systems-level strategy combining cutting-edge materials for selective filtering and catalysis, real-time monitoring systems, and ad+aptive control techniques is needed to address these issues. To improve the dependability, effectiveness, and economic viability of fuel cell-based WtE technologies for industrial applications, ongoing research in integrated WtE systems is concentrating on optimising these interfaces.

5.5 WASTE-TO-ENERGY SYSTEMS INTEGRATION

5.5.1 System Design and Integration of Waste-to-Energy with Fuel Cells

A multi-step process is required to design and integrate a WtE system with fuel cell technology, guaranteeing the smooth conversion of municipal or industrial waste into usable heat and electricity [101]. The main concept is to connect a suitable fuel cell system that can effectively turn the resulting fuel (hydrogen or syngas) into electricity to a waste conversion unit, such as a gasifier, pyrolysis reactor, anaerobic digester, or reformer [102, 103]. Aligning the waste conversion process's operational, chemical, and physical properties with the fuel cell's specifications is essential for a successful integration.

For instance, in a gasification-based WtE system, syngas—a blend of CO, H_2, CH_4, and other gases—is first produced from organic industrial waste like biomass or plastic [104, 105]. To get rid of contaminants like tar, sulphur compounds, particulates, and chlorides, this syngas must next be cleaned and conditioned. Following purification, the syngas can be fed into an SOFC, which is ideal for mixed gas feeds because of its high operating temperature and fuel flexibility. Known as cogeneration or CHP, this setup also allows heat from the SOFC to be recycled back into the gasifier, increasing overall energy efficiency [106, 107].

As an illustration of biological integration, waste from the food industry is processed in an anaerobic digester to create biogas, which is mostly made up of CO_2 and methane [108]. By eliminating CO_2 and sulphur compounds, this biogas can be refined to produce hydrogen or methane that is almost pure. The methane-rich gas can then be directly used by an MCFC, producing heat and electricity with low emissions [109]. In these situations, integration also entails setting up compressors, heat exchangers, and gas separation membranes to manage gas flows, maximise system efficiency, and recover waste heat.

Scale, feedstock variability, emission limits, and site-specific energy requirements must all be considered when designing a system. Heat recovery systems, gas handling and cleaning modules, process control units, and safety features are usually included. Furthermore, modularity is crucial, particularly in industrial environments where waste production rates can fluctuate. Modular systems enable flexible operation and gradual capacity expansion.

Maintaining constant fuel quality, reducing system downtime, reaching high overall system efficiency, and guaranteeing cost-effectiveness are some of the main integration challenges. To get around these, engineers frequently employ sophisticated control systems that continuously monitor temperature, load, and gas composition to modify the waste processor and fuel cell's operations.

In summary, careful consideration of waste conversion pathways, fuel compatibility, gas conditioning, and thermal integration is necessary when integrating WtE systems with fuel cells. When implemented correctly, these systems provide a sustainable, effective, and clean means of turning waste into useful energy, particularly in industrial environments with consistent waste streams.

5.5.2 Energy Recovery and Power Generation Efficiency

When assessing the effectiveness of WtE systems combined with fuel cell technologies, energy recovery and power generation efficiency are crucial factors. With the least possible impact on the environment, these systems seek to optimise the transformation of chemical energy stored in waste materials into electrical and thermal energy that can be used.

In this regard, fuel cells are clearly superior to traditional combustion-based power systems because of their inherent high electrical efficiency [110]. For example, depending on the type and fuel quality, fuel cells can reach electrical efficiencies of 40–60%, whereas conventional steam turbines or internal combustion engines used in traditional WtE plants may only reach electrical efficiencies in the range of 20–35% (Table 5.2). Because the waste heat produced by the fuel cell is captured and used for industrial processes, space heating, or preheating feedstock in the waste conversion units, CHP configurations can achieve overall system efficiencies of over 80%.

Syngas generated from organic and municipal wastes is utilised as a feedstock for fuel cells such as SOFCs or MCFCs in thermal conversion processes like gasification or pyrolysis (Table 5.2). Because these fuel cells run at high temperatures, hydrocarbons can reform internally, and syngas can be used directly. Additionally, efficient waste heat recovery is made possible by the high operating temperature, which greatly enhances system efficiency.

Biogas produced by biological conversion processes, like anaerobic digestion, is mostly made up of methane. This biogas can be purified and used as an effective fuel for MCFCs or PEMFCs (Table 5.2). PEMFCs respond rapidly to changes in load and are appropriate for decentralised or smaller-scale applications where reliable biogas production is available, despite operating at lower temperatures and having a slightly lower electrical efficiency.

In addition to the fuel cell itself, other factors that affect efficiency in WtE systems include the quality of the feedstock, the degree of gas cleaning and conditioning, and the efficiency of the upstream conversion process. Sulphur, tars, and particulates are examples of impurities that can shorten the lifespan and performance of fuel cells, which lowers total energy recovery [111].

The energy recovery process is further improved by the clever integration of thermal insulation, waste heat recovery systems, and energy management software [112]. WtE systems with fuel cell integration can outperform conventional energy conversion techniques in terms of energy recovery and environmental performance by maximising fuel utilisation and minimising energy losses at every stage.

In summary, the incorporation of fuel cells into WtE frameworks offers high power generation efficiency, efficient waste reduction, and the possibility of sustainable operation in urban and industrial environments, making it a significant advancement in clean energy production.

5.5.3 Waste Handling, Fuel Processing, and Storage

Fundamental elements of the design and functioning of WtE systems integrated with fuel cell technologies include waste handling, fuel processing, and storage. By taking these precautions, waste feedstocks are consistently transformed into clean, fuel cell-compatible inputs while preserving environmental compliance, operational effectiveness, and safety.

The collection and segregation stage is where efficient waste management starts. It is necessary to identify and separate various waste types in industrial settings, such as organic, plastic, chemical, or sludge, based on their compatibility with downstream processes and potential for energy recovery. For example, solid or carbon-rich wastes are better suited for thermal processes like pyrolysis and gasification, whereas organic and biodegradable wastes are better suited for biological conversion techniques like anaerobic digestion [113]. Effective preprocessing methods, like shredding, drying, or dewatering, are used to lower the waste's volume, increase its uniformity, and boost its energy density [114, 115].

The pretreated waste is subsequently transformed into liquid or gaseous fuels appropriate for fuel cell operation through fuel processing [116]. Syngas, which is mostly composed of hydrogen (H_2), carbon monoxide (CO), and methane (CH_4), is produced by gasification of carbonaceous materials in thermal routes [117, 118]. To prevent tar, particulates, sulphur compounds, and other impurities from poisoning the catalyst layers in fuel cells, this syngas needs to be cleaned [106, 119]. Similarly, before being fed to fuel cells such as PEMFCs or MCFCs, biogas from anaerobic digestion is purified to eliminate moisture, carbon dioxide (CO_2), and hydrogen sulphide (H_2S) [120, 121].

Clean hydrogen is separated from waste-derived hydrocarbons for use in chemical processing methods like partial oxidation or steam reforming. These reforming procedures are particularly important for low-temperature fuel cells, which need high-purity hydrogen [122].

Storage is an important factor, especially when it comes to guaranteeing a steady and uninterrupted fuel supply for the system. To preserve quality and stop leaks, syngas and biogas are usually kept under pressure in sealed vessels [123]. Depending on the facility's size and safety regulations, hydrogen that is used directly must be stored as compressed gas, cryogenic liquid, or absorbed in solid-state materials like metal hydrides [124, 125]. In every situation, the right safety precautions must be taken to control the risks of gas storage, such as material compatibility, pressure buildup, and explosion hazards.

To guarantee a steady supply of high-quality fuel to the fuel cell stack, integrated WtE fuel cell systems require coordination between the waste supply, fuel processing capability, and storage infrastructure. This dynamic is frequently managed by automation and monitoring systems, which modify preprocessing and storage operations in real time in response to energy demand and waste input variability.

In summary, the effective implementation of WtE systems utilising fuel cells depends on efficient waste management, accurate fuel processing, and dependable storage. These actions promote safe and sustainable energy production in industrial settings while also improving system longevity and energy conversion efficiency.

5.5.4 Combined Heat and Power Potential in Industrial Processes

When combined with fuel cell-based WtE systems, CHP, commonly referred to as cogeneration, offers a substantial chance to improve energy efficiency in industrial processes [126]. CHP significantly lowers energy losses that would otherwise arise from distinct generating processes by generating electricity and useful thermal energy simultaneously from a single fuel source [127].

A large portion of the energy produced by burning fuel is wasted as waste heat in conventional power plants. Nevertheless, depending on the type and configuration of the fuel cell, this heat can be captured and used for industrial heating, drying, or steam generation in a fuel cell-based CHP system, resulting in overall system efficiencies of 70–90% [128, 129]. Because of their high operating temperatures, which allow for the direct use of the heat generated during electrochemical reactions, SOFCs and MCFCs are especially well-suited for CHP.

Paper and pulp, textile production, food processing, chemical manufacturing, and other industrial sectors frequently need both thermal and electrical energy at the same time [130]. They are therefore perfect candidates for fuel cell-CHP integration. For example, a food processing facility can use the electricity to run its conveyor and refrigeration systems and use the recovered heat to cook or pasteurise food [131, 132].

Furthermore, by combining CHP and WtE, businesses can create a closed-loop energy system by using their own waste as feedstock to produce heat and power, such as volatile compounds, wastewater sludge, or organic byproducts [133, 134]. In addition to increasing resilience and energy independence, this lessens dependency on fossil fuels and lowers greenhouse gas emissions.

Distributed CHP systems can also help stabilise local power networks and supply backup power during grid outages, which has reliability and financial advantages [71]. CHP adoption in industrial settings is further encouraged by incentives offered by governments or carbon credit programmes [135, 136].

In WtE applications, the combination of fuel cell technology and CHP has the potential to revolutionise sustainable industrial energy systems by maximising resource use and supporting global decarbonisation objectives.

5.6 CASE STUDIES AND INDUSTRIAL APPLICATIONS

5.6.1 Real-World Examples of Waste-to-Energy Fuel Cell Deployments

Real-world implementations of WtE systems that integrate fuel cell technologies show promising results, as illustrated in Table 5.2. For example, anaerobic digestion in dual-fuel engines yielded 51% brake thermal efficiency [137], MFCs removed 77% of COD from wastewater with a power density of 175.7 W/m^2 [138], and SOFCs assisted gasification of plastic waste using artificial intelligence (AI). Large-scale incineration-based WtE in China increased energy recovery from 6.18% to 63.67% even without the use of fuel cells [139].

5.6.2 Pilot-Scale and Commercial Projects in Various Industries

Across a range of industries, pilot-scale and commercial WtE projects incorporating fuel cell technologies are progressing, showcasing the potential for sustainable energy solutions. These programmes demonstrate a variety of uses and advantages, ranging from municipal waste management to industrial manufacturing.

5.6.2.1 The Carbon Capture Pilot Plant of ExxonMobil and Fuel Cell Energy

Carbonate fuel cell technology is being used by ExxonMobil and FuelCell Energy to develop a pilot carbon capture and storage (CCS) plant. By capturing carbon dioxide

TABLE 5.2
Real-world examples of WtE fuel cell deployments

Research Work	Conversion Techniques	Type of Fuel Cells	Performance Metrics	Reference
AI-assisted gasification of plastic waste using SOFCs	Plastic waste gasification	Solid Oxide Fuel Cell (SOFC)	Optimal efficiency: 41.89 and 860.71 kg/MWh achieved	[140]
Microbial fuel cells to treat wastewater	Biological conversion	Microbial fuel cell (MFC)	77% COD removal efficiency with a power density of 175.7 W/m^2	[138]
Energy district optimisation for FCEVs	Integration of waste and recycling fleet	Fuel cell electric vehicles (FCEVs)	System efficiency is about 50%	[141]
Biogas utilisation in dual-fuel engines	Anaerobic digestion of food and animal waste	Dual-fuel engine (biogas and diesel)	In contrast to CO (17%) and BSFC (8%), BTE (51%) and NOx (24%) emerged as the dominant parameters.	[137]
Waste-to-energy, example in China	Incineration of municipal solid waste	Steam turbines (not fuel cells)	WtE increased from 6.18% to 63.67%	[139]
Microbial fuel cells from orange waste	Biological conversion	MFC	Voltage output of 357 mV	[142]
SOFC from weed waste	In situ gasification	SOFC	Power and energy density are respectively 0.41 Wcm^{-2} and 0.42 Whg^{-1}	[143]
Waste-to-biofuel production	Hydrothermal liquefaction (HTL), transesterification, and incineration	Biofuel	HTL of net present value of ~$200 million, transesterification produced 2,850,000 MJ/year, and Incineration emitting ~3000 tonnes/year of CO_2.	[144]
Waste-to-hydrogen (WtH) systems for electric bus	Gasification and fermentation	Hydrogen systems	Reduced CO_2-eq emissions per kg H_2 by 50–69% as compared to steam methane reforming.	[145]
Taguchi-multi-response assisted waste-to-energy density	Enzymatic hydrolysis of lignin	Energy storage (e.g., capacitor)	pH with 33.30% of energy storage enhancement	[44]

emissions from industrial sources before their release into the atmosphere, this facility may provide cost-effective decarbonisation solutions for a range of industries [146].

5.6.2.2 BMW's Gas-to-Hydrogen Landfill Project

A pilot project to turn landfill gas into hydrogen to power more than 300 material handling equipment (MHE) units has been put into place at BMW's assembly plant in South Carolina. This project shows that producing hydrogen on-site from waste materials for industrial uses is feasible [147].

5.6.2.3 California Wood Waste Gasification Project by Mote Inc.

To achieve negative emissions, Mote Inc. is integrating carbon sequestration into a pilot project in Los Angeles that gasifies wood waste to produce hydrogen. The potential of integrating carbon capture technologies with WtE processes is demonstrated by this project [148].

5.6.2.4 Australia's Yarwun Biofuels Pilot Plant in Queensland

A biofuels pilot plant in central Queensland is using a chemical looping process to create hydrogen from waste. To support refining operations and show how waste-derived hydrogen can be integrated into industrial processes, the excess hydrogen powers a 200–400 kW fuel cell [149].

5.6.2.5 Ghana's Ashaiman Waste-to-Energy Plant

Ghana's Ashaiman plant produces about 580 MWh annually by treating 45 tonnes of organic and faecal waste per day. This facility serves as an example of how WtE technologies can be used to meet energy and waste management demands in developing nations [150].

These initiatives demonstrate the adaptability and scalability of incorporating fuel cell technologies into WtE systems in various industries and geographical areas. They provide real-world examples of how businesses can use waste materials to generate energy, promoting both energy efficiency and environmental sustainability.

5.6.3 Lessons Learned and Best Practices

Real-world examples of WtE systems combined with fuel cell technologies have taught us how crucial consistent feedstock, thorough pretreatment, and system integration are to achieving the best results. To avoid catalyst degradation and increase long-term efficiency, projects involving gasification and SOFCs highlight the necessity of exact control over syngas composition and impurity removal. For biological pathways like anaerobic digestion and MFCs, stable microbial populations and ideal environmental conditions are essential for long-term waste degradation and electricity production. Best practices include using digital tools like AI and machine learning to improve process control and predictive maintenance, adopting hybrid configurations (such as fuel cell–CHP systems) for maximum energy recovery, and designing modular systems that enable scalability [151]. Additionally, pilot-to-commercial transitions have been fueled by economic incentives, policy alignment, and stakeholder engagement.

All these observations suggest that to guarantee the success and reproducibility of fuel cell-based WtE projects, customised solutions that consider the features of industrial waste, regulatory frameworks, and end-use energy requirements are required.

5.7 TECHNO-ECONOMIC AND ENVIRONMENTAL ASSESSMENT

Significant benefits and noteworthy challenges are revealed by a thorough techno-economic and environmental evaluation of WtE systems combined with fuel cell technologies. Depending on the fuel cell technology and system configuration, the estimated capital expenditures for such systems range from $3,000 to $10,000 per kilowatt. Higher electrical efficiencies and fewer requirements for pollutant treatment, however, result in relatively lower operating costs. For example, compared to traditional combustion-based WtE technologies, SOFCs and MCFCs can achieve electrical efficiencies exceeding 50% [152].

WtE fuel cell systems usually have a payback period of 5–12 years, depending on variables like scale, feedstock availability, and energy market incentives. Fuel cell-based WtE systems generate fewer greenhouse gas emissions than conventional incineration or landfill gas combustion, according to life cycle assessments (LCAs). For instance, combining WtE plants with CCS can further lower carbon footprints and possibly lead to negative emissions [153].

Fuel cells provide better energy conversion efficiencies, fuel flexibility, and the added benefit of CHP capabilities when compared to traditional energy recovery techniques like internal combustion engines and steam turbines. These characteristics improve overall system utility and ROI, especially in industrial settings where waste heat can be efficiently used.

Notwithstanding these advantages, there are still drawbacks, such as high upfront capital expenses and the requirement for clean, conditioned fuel to stop system deterioration. The commercial viability of WtE-fuel cell technologies is being enhanced by advancements in fuel flexibility, modularity, and system optimisation. As a result, these technologies hold promise as a part of sustainable energy infrastructure.

5.8 REVIEW OF IOT APPLICATIONS IN WASTE-TO-ENERGY SYSTEMS

The integration of IoT technologies into WtE systems has gained momentum in recent years, offering new dimensions of efficiency, automation, and intelligence across the energy conversion process. Numerous studies have demonstrated that IoT can significantly enhance WtE system performance by enabling real-time data acquisition, predictive analytics, and decentralised control [154–157].

IoT applications in WtE systems typically involve sensor networks, cloud-based platforms, edge computing, and machine learning algorithms. These technologies are used to monitor critical parameters such as waste composition, feedstock moisture content, temperature, pressure, gas output quality, and fuel cell performance metrics [158–160]. For example, in smart pyrolysis and gasification units, IoT-enabled sensors have been deployed to dynamically adjust operational parameters in response to fluctuations in waste input, thereby optimising syngas quality and energy yield [160, 161].

In fuel cell-based WtE systems, IoT plays a crucial role in ensuring consistent performance and operational safety. Researchers have implemented IoT frameworks for remote diagnostics and predictive maintenance of solid oxide and PEM fuel cells. Such systems can identify anomalies in voltage, temperature, and fuel supply, allowing for proactive adjustments before failures occur. This predictive capability reduces downtime and improves system longevity, which is critical in industrial environments [162–164].

Additionally, IoT facilitates smart waste classification and sorting prior to conversion, leveraging image recognition and AI to segregate recyclable, organic, and hazardous materials. This enhances the quality and consistency of feedstock used for energy generation, particularly in thermal and biochemical conversion routes [8, 165].

Emerging research also highlights the role of IoT in integrating WtE plants with smart grids and industrial energy management systems. By transmitting real-time data on energy output and waste input, WtE systems can better align with demand-response strategies and optimise power distribution within an industrial setting [10, 166, 167].

Despite these advancements, challenges remain, including data security, interoperability between devices, and the need for standardised protocols. However, ongoing research continues to address these issues through blockchain integration, open-source IoT platforms, and edge AI solutions [168, 169].

In summary, the application of IoT in WtE systems enhances automation, system intelligence, and operational efficiency. It provides a foundation for more resilient, adaptive, and sustainable energy systems aligned with Industry 4.0 and circular economy principles.

5.9 CHALLENGES, OPPORTUNITIES, AND FUTURE DIRECTIONS

One promising path towards the production of sustainable energy is the incorporation of fuel cell technologies into WtE systems. To reach their full potential, several opportunities and challenges must be resolved.

Fuel impurities are one of the main technical issues with WtE fuel cell systems since they can cause catalyst poisoning and fuel cell component deterioration. For example, impurities like sulphur compounds and particulates in waste gasification-derived syngas can negatively impact the efficiency of MCFCs and SOFCs. Furthermore, the scalability of these systems is still an issue because more research and development are needed to maintain durability and efficiency over longer time periods and larger scales.

Barriers related to infrastructure, policy, and regulations also prevent WtE fuel cell technologies from being widely used. The integration of cutting-edge technologies like fuel cells is hampered by the fact that many regions' current energy policies and regulations are designed to accommodate traditional energy systems. Significant obstacles are also presented by the absence of a standardised infrastructure for the production, distribution, and storage of hydrogen [170].

The development of new materials and catalysts is essential to improving fuel cell system performance and lowering costs. To increase the oxygen reduction reaction

(ORR) and overall efficiency of MFCs, recent research has concentrated on creating substitute catalysts to replace pricey platinum-based materials. Furthermore, studies into strong and affordable membrane materials are being conducted to address concerns about the performance and longevity of fuel cells [26, 171].

Promising answers to some of the issues WtE fuel cell technologies face can be found in hybrid system innovations and the use of AI for process control. The efficiency and dependability of a system can be improved by hybrid systems that integrate fuel cells with additional renewable energy sources, like solar or wind. Furthermore, AI-based control strategies can enhance these systems' flexibility in response to changing operating conditions, forecast maintenance requirements, and optimise energy management.

In the future, WtE fuel cell technologies could be very important for industrial decarbonisation and the circular economy. These systems can lessen dependency on fossil fuels and greenhouse gas emissions by turning waste into useful energy. The feasibility of these technologies is anticipated to be further improved by emerging trends in fuel cell development, such as the investigation of novel materials and system configurations. More robust and efficient energy networks can also be made possible by integration with smart grids and renewable energy sources.

Policymakers and industry stakeholders should think about putting supportive laws into place, funding the construction of infrastructure, and encouraging cooperative research projects to promote the development of WtE fuel cell technologies. These technologies can be expanded to offer efficient and sustainable energy solutions in the future by tackling the technical, legal, and financial obstacles.

This chapter concludes that integrating fuel cell technologies with IoT presents a promising pathway for enhancing industrial WtE systems. Fuel cells offer high efficiency, fuel flexibility, and low emissions, making them well-suited for converting diverse industrial waste streams into clean energy. When paired with IoT, these systems gain improved monitoring, control, and optimisation capabilities. Together, these innovations address key challenges related to energy efficiency, scalability, and sustainability, positioning WtE as a viable solution for low-carbon industrial development and circular economy goals.

REFERENCES

1. Goldthau, A., *Governing global energy: Existing approaches and discourses.* Current Opinion in Environmental Sustainability, 2011. 3(4): p. 213–217.
2. Cook, M., *Trends in global energy supply and demand*, in *Developments in Petroleum Science.* 2021, Elsevier. p. 15–42.
3. Obada, D., et al., *Mechanical properties of natural hydroxyapatite using low cold compaction pressure: Effect of sintering temperature.* Materials Chemistry and Physics, 2020. 239: p. 122099.
4. Abifarin, J., et al., *Experimental data on the characterization of hydroxyapatite synthesized from biowastes.* Data in Brief, 2019. 26: p. 104485.
5. Abifarin, J.K., et al., *Fabrication of mechanically enhanced hydroxyapatite scaffold with the assistance of numerical analysis.* The International Journal of Advanced Manufacturing Technology, 2021: 118: p. 3331–3344.
6. Nemerow, N.L., *Industrial collaborative solutions*, in *Environmental Solutions.* 2005, Elsevier. p. 249–295.

7. Gaikwad, N. and G.S. Gund, *Future aspects of machine learning/automation for the waste management*, in *Harnessing Automation and Machine Learning for Resource Recovery and Value Creation*. 2025, Elsevier. p. 461–474.
8. Lakhouit, A., *Revolutionizing urban solid waste management with AI and IoT: A review of smart solutions for waste collection, sorting, and recycling*. Results in Engineering, 2025. **25**: p. 104018.
9. Cheah, C.G., et al., *Innovation designs of industry 4.0 based solid waste management: Machinery and digital circular economy*. Environmental Research, 2022. 213: p. 113619.
10. Mishra, P. and G. Singh, *Energy management systems in sustainable smart cities based on the internet of energy: A technical review*. Energies, 2023. 16(19): p. 6903.
11. Abifarin, F.B., Z. Musa, and J.K. Abifarin, *Mechanical processing of hydroxyapatite through sintering and multi-objective optimization technique for biomedical application*. MRS Advances, 2023. 8(9): p. 532–537.
12. Abifarin, J.K., *Taguchi grey relational analysis on the mechanical properties of natural hydroxyapatite: Effect of sintering parameters*. The International Journal of Advanced Manufacturing Technology, 2021. 117(1): p. 49–57.
13. Obada, D., et al., *Mechanical measurements of pure and kaolin reinforced hydroxyapatite-derived scaffolds: a comparative study*. Materials Today: Proceedings, 2021. 38: p. 2295–2300.
14. Abifarin, J.K., et al., *Taguchi grey relational optimization of the multi-mechanical characteristics of kaolin reinforced hydroxyapatite: effect of fabrication parameters*. International Journal of Grey Systems, 2021. 1(2): p. 20–32.
15. Oparanti, S., et al., *Synthesis and characterization of cooling biodegradable nanofluids from non-edible oil for high voltage application*. Materials Chemistry and Physics, 2022. 277: p. 125485.
16. Joseph, S., et al., *Sustainable synthesis of shea butter-derived bio-lubricants: A green alternative to mineral oils*. Next Sustainability, 2025. 5: p. 100128.
17. Abifarin, J.K. and O.A. Owolabi, *New insight to the mechanical reliability of porous and nonporous hydroxyapatite*. Journal of the Australian Ceramic Society, 2023. 59(1): p. 43–55.
18. Saxena, V., *Water quality, air pollution, and climate change: Investigating the environmental impacts of industrialization and urbanization*. Water, Air, & Soil Pollution, 2025. 236(2): p. 1–40.
19. Abifarin, J.K., J.F. Torres, and Y. Lu, *2D materials for enabling hydrogen as an energy vector*. Nano Energy, 2024: p. 109997.
20. Abifarin, J.K., et al., *Inhibitor additives to mitigate fossil fuel emissions and its potential role in promoting the energy transition in global cities*. Energy, Ecology and Environment, 2025. 10: p. 411–436.
21. Hsu, H.-W., et al., *Toward sustainability of Waste-to-Energy: An overview*. Energy Conversion and Management, 2024. 321: p. 119063.
22. Shi, Y., Y. Huang, and J. Xu, *A clean optimization approach for sustainable waste-to-energy using integrated technology*. Renewable Energy, 2024. 221: p. 119750.
23. Abifarin, J.K. and F.B. Abifarin, *Carbon emission reduction and hydrogen production maximization from carbon emission-based hydrogen sources*. Environmental Science and Pollution Research, 2024. 32: p. 28489–28501.
24. Slate, A.J., et al., *Microbial fuel cells: An overview of current technology*. Renewable and Sustainable Energy Reviews, 2019. 101: p. 60–81.
25. Ieropoulos, I.A., et al., *Comparative study of three types of microbial fuel cell*. Enzyme and Microbial Technology, 2005. 37(2): p. 238–245.
26. Chhachhiya, N., et al., *Transformative potential of optimized microbial fuel cell designs and materials for eco-friendly management of hazardous chemical waste*. Journal of Water Process Engineering, 2025. 69: p. 106647.

27. Hamedani, E.A., A. Abasalt, and S. Talebi, *Application of microbial fuel cells in wastewater treatment and green energy production: A comprehensive review of technology fundamentals and challenges.* Fuel, 2024. 370: p. 131855.
28. Badwal, S.P., et al., *Emerging electrochemical energy conversion and storage technologies.* Frontiers in Chemistry, 2014. 2: p. 79.
29. Srivastava, R.K., et al., *Sustainable energy from waste organic matters via efficient microbial processes.* Science of the Total Environment, 2020. 722: p. 137927.
30. Bilal, M., D. Niu, and Z. Wang, *Biotechnological strategies and perspectives for food waste treatment: The role of lactic acid and microbial biomass.* Waste and Biomass Valorization, 2025. 16(2): p. 547–569.
31. Arena, U., *Process and technological aspects of municipal solid waste gasification. A review.* Waste Management, 2012. 32(4): p. 625–639.
32. Saritha, G., *Biomedical waste management.* In Epidemiology and Environmental Hygiene in Veterinary Public Health (ed) Rana, T., 2025: p. 523–535. John Wiley & Sons, Inc, Print ISBN:9781394208159, DOI: https://doi.org/10.1002/9781394208180.ch45.
33. Kassawe, A. and E. Getahun, *Solid waste generation rate, composition analysis, and proposed management plan: A case study of main market centers of bahir dar city.* The Scientific World Journal, 2025. 2025(1): p. 8504268.
34. Artiola, J., *Industrial waste and municipal solid waste treatment and disposal*, in *Environmental and Pollution Science.* 2019, Elsevier. p. 377–391.
35. Saravanan, A., et al., *Effective water/wastewater treatment methodologies for toxic pollutants removal: Processes and applications towards sustainable development.* Chemosphere, 2021. 280: p. 130595.
36. Munsif, R., et al., *Industrial air emission pollution: Potential sources and sustainable mitigation*, in *Environmental Emissions.* 2021, IntechOpen. p. 13. DOI: 10.5772/intechopen.93104.
37. Abifarin, J.K., F.B. Abifarin, and J.C. Ofodu, *Composite catalysts for enabling fossil fuel emissions capture.* Science of the Total Environment, 2025. 1000: p. 180379.
38. Manisalidis, I., et al., *Environmental and health impacts of air pollution: A review.* Frontiers in Public Health, 2020. 8: p. 14.
39. Hill, C.T., *Thermal pollution and its control. Environmental Affairs*, 1972. 2: p. 406.
40. Kumar, M. and G. Mishra, *Causes and impacts of water pollution on various water bodies in the state of Rajasthan, India: A review. Environment and Ecology*, 2024. 42: p. 645–654.
41. Trivedi, K., et al., *Assessment and characterization of solid and hazardous waste from inorganic chemical industry: Potential for energy recovery and environmental sustainability.* Journal of Environmental Management, 2024. 367: p. 122036.
42. Elehinafe, F.B., et al., *Separation processes for the treatment of industrial flue gases–Effective methods for global industrial air pollution control.* Heliyon, 2024. **10**(11). Paper ID: e32428.
43. Jafarinejad, S., *Control and treatment of sulfur compounds specially sulfur oxides (SOx) emissions from the petroleum industry: A review.* Chemistry International, 2016. 2(4): p. 242–253.
44. Abifarin, J.K., F.B. Abifarin, and J.C. Ofodu, *Lignin-hydrothermal fabrication of 3D hierarchical porous carbon for energy storage application.* Bioresource Technology Reports, 2024. 26: p. 101883.
45. Appels, L., et al., *Anaerobic digestion in global bio-energy production: Potential and research challenges.* Renewable and Sustainable Energy Reviews, 2011. 15(9): p. 4295–4301.
46. Kiyasudeen S, K., et al., *An introduction to anaerobic digestion of organic wastes*, in *Prospects of Organic Waste Management and the Significance of Earthworms*, 2016: p. 23–44. Springer, Cham.

47. Sumathi, S. and Y.-T. Hung, *Treatment of pulp and paper mill wastes*, in *Handbook of Industrial and Hazardous Wastes Treatment*. 2004, CRC Press. p. 507–555.
48. Gupta, A. and R. Gupta, *Treatment and recycling of wastewater from pulp and paper mill*, in *Advances in Biological Treatment of Industrial Waste Water and Their Recycling for a Sustainable Future*. 2018, Springer. p. 13–49.
49. Beyene, H.D., A.A. Werkneh, and T.G. Ambaye, *Current updates on waste to energy (WtE) technologies: a review*. Renewable Energy Focus, 2018. 24: p. 1–11.
50. Khan, A.H., et al., *Municipal solid waste generation and the current state of waste-to-energy potential: State of art review*. Energy Conversion and Management, 2022. 267: p. 115905.
51. Hassan, M., et al., *Current challenges and future perspectives associated with configuration of microbial fuel cell for simultaneous energy generation and wastewater treatment*. International Journal of Hydrogen Energy, 2024. 50: p. 323–350.
52. Logan, B.E., et al., *Microbial fuel cells: Methodology and technology*. Environmental Science & Technology, 2006. 40(17): p. 5181–5192.
53. Hall, J. and R. Kerr, *Innovation dynamics and environmental technologies: The emergence of fuel cell technology*. Journal of Cleaner Production, 2003. 11(4): p. 459–471.
54. Serfass, J.A., M.K. Bergman, and W. Rodenhiser, *Commercial, environmental and legislative factors that influence the implementation of fuel cells*. Journal of Power Sources, 1994. 49(1–3): p. 193–208.
55. Lee, Y., Y.J. Kim, and M.C. Lee, *Navigating stakeholder perspectives on hydrogen generation technologies: A Q-methodology study of current hydrogen production policy in South Korea*. International Journal of Hydrogen Energy, 2025. 113: p. 777–786.
56. Fang, T., A. von Jouanne, and A. Yokochi, *Review of electrochemical systems for grid scale power generation and conversion: Low-and high-temperature fuel cells and electrolysis processes*. Energies, 2025. 18(10): p. 2493.
57. Sharma, V., et al., *Applications of solid waste-derived carbon nanomaterials in fuel cell*, in *Waste Derived Carbon Nanomaterials. Volume* 2. 2025, ACS Publications. p. 103–121.
58. Singh, R., et al., *Advances in additive manufacturing of fuel cells: A review of technologies, materials, and challenges*. Sustainable Materials and Technologies, 2025. 43: p. e01317. https://doi.org/10.1016/j.susmat.2025.e01317
59. Simya, O., et al., *Engineered nanomaterials for energy applications*, in *Handbook of Nanomaterials for Industrial Applications*, 2018: Elsevier. p. 751–767.
60. Sharaf, O.Z. and M.F. Orhan, *An overview of fuel cell technology: Fundamentals and applications*. Renewable and Sustainable Energy Reviews, 2014. 32: p. 810–853.
61. Chandan, A., et al., *High temperature (HT) polymer electrolyte membrane fuel cells (PEMFC)–A review*. Journal of Power Sources, 2013. 231: p. 264–278.
62. Li, Q., et al., *Approaches and recent development of polymer electrolyte membranes for fuel cells operating above* 100 *C*. Chemistry of Materials, 2003. 15(26): p. 4896–4915.
63. Chuahy, F.D. and S.L. Kokjohn, *Solid oxide fuel cell and advanced combustion engine combined cycle: A pathway to 70% electrical efficiency*. Applied Energy, 2019. 235: p. 391–408.
64. Cui, C., et al., *Review of molten carbonate-based direct carbon fuel cells*. Materials for Renewable and Sustainable Energy, 2021. 10: p. 1–24.
65. Lin, B.Y., D.W. Kirk, and S.J. Thorpe, *Performance of alkaline fuel cells: A possible future energy system?* Journal of Power Sources, 2006. 161(1): p. 474–483.
66. Zhang, H., G. Lin, and J. Chen, *The performance analysis and multi-objective optimization of a typical alkaline fuel cell*. Energy, 2011. 36(7): p. 4327–4332.
67. Acar, C., A. Beskese, and G.T. Temur, *Comparative fuel cell sustainability assessment with a novel approach*. International Journal of Hydrogen Energy, 2022. 47(1): p. 575–594.

68. Tomczyk, P., *MCFC versus other fuel cells—Characteristics, technologies and prospects*. Journal of Power Sources, 2006. 160(2): p. 858–862.
69. Singhal, S.C. and K. Kendall, *High-Temperature Solid Oxide Fuel Cells: Fundamentals, Design and Applications*. 2003, Elsevier.
70. Stambouli, A.B. and E. Traversa, *Fuel cells, an alternative to standard sources of energy*. Renewable and Sustainable Energy Reviews, 2002. 6(3): p. 295–304.
71. Baroutaji, A., et al., *PEMFC poly-generation systems: Developments, merits, and challenges*. Sustainability, 2021. 13(21): p. 11696.
72. Rosli, R., et al., *A review of high-temperature proton exchange membrane fuel cell (HT-PEMFC) system*. International Journal of Hydrogen Energy, 2017. 42(14): p. 9293–9314.
73. Campanari, S., G. Manzolini, and P. Chiesa, *Using MCFC for high efficiency CO2 capture from natural gas combined cycles: Comparison of internal and external reforming*. Applied Energy, 2013. 112: p. 772–783.
74. Abdelkareem, M.A., et al., *Fuel cells for carbon capture applications*. Science of The Total Environment, 2021. 769: p. 144243.
75. Wang, H.-S., et al., *A high-yield and ultra-low-temperature methanol reformer integratable with phosphoric acid fuel cell (PAFC)*. Energy, 2017. 133: p. 1142–1152.
76. Aziz, M., A. Darmawan, and F.B. Juangsa, *Hydrogen production from biomasses and wastes: A technological review*. International Journal of Hydrogen Energy, 2021. 46(68): p. 33756–33781.
77. Malik, F.R., et al., *Overview of hydrogen production technologies for fuel cell utilization*. Engineering Science and Technology, an International Journal, 2023. 43: p. 101452.
78. Whitty, K.J., H.R. Zhang, and E.G. Eddings, *Emissions from syngas combustion*. Combustion Science and Technology, 2008. 180(6): p. 1117–1136.
79. Razzak, S.A., *Municipal solid and plastic waste derived high-performance biochar production: A comprehensive review*. Journal of Analytical and Applied Pyrolysis, 2024. **181**: p. 106622.
80. Callegari, A. and A.G. Capodaglio, *Properties and beneficial uses of (bio) chars, with special attention to products from sewage sludge pyrolysis*. Resources, 2018. 7(1): p. 20.
81. Demirbas, A., et al., *Aerobic digestion of sewage sludge for waste treatment*. Energy Sources, Part A: Recovery, Utilization, and Environmental Effects, 2017. 39(10): p. 1056–1062.
82. Alves, H.J., et al., *Overview of hydrogen production technologies from biogas and the applications in fuel cells*. International Journal of Hydrogen Energy, 2013. 38(13): p. 5215–5225.
83. Saadabadi, S.A., et al., *Solid Oxide Fuel Cells fuelled with biogas: Potential and constraints*. Renewable Energy, 2019. 134: p. 194–214.
84. Iulianelli, A., et al., *Advances on methane steam reforming to produce hydrogen through membrane reactors technology: A review*. Catalysis Reviews, 2016. 58(1): p. 1–35.
85. Hamad, T.A., et al., *Study of combined heat, hydrogen and power system based on a molten carbonate fuel cell fed by biogas produced by anaerobic digestion*. Energy Conversion and Management, 2014. 81: p. 184–191.
86. Archer, S.A. and R. Steinberger-Wilckens, *Systematic analysis of biomass derived fuels for fuel cells*. International Journal of Hydrogen Energy, 2018. 43(52): p. 23178–23192.
87. Sharma, M., D. Bose, and T. Banerjee, *Low carbon energy system: Role of fuel cell technology*, in *Low Carbon Energy Supply Technologies and Systems*. 2020, CRC Press. p. 123–142.
88. Zhang, S., et al., *Design and research of a novel solid oxide fuel cell with thermal energy storage for load tracking*. Energy Conversion and Management, 2023. 294: p. 117560.

89. Rasi, S., et al., *Cascade processing of softwood bark with hot water extraction, pyrolysis and anaerobic digestion.* Bioresource Technology, 2019. 292: p. 121893.
90. Deb, N., et al., *Efficient technology for food waste valorization: An integrated anaerobic digestion—microbial electrolysis cell for biomethane production.* International Journal of Environmental Science and Technology, 2025. 22(7): p. 5609–5622.
91. Chen, C.-H., C.-T. Yu, and W.-H. Chen, *Improvement of steam methane reforming via in-situ CO2 sorption over a nickel-calcium composite catalyst.* International Journal of Hydrogen Energy, 2021. 46(31): p. 16655–16666.
92. Khosravani, H., H.R. Rahimpour, and M.R. Rahimpour, *Characteristics of syngas impurities; Physical and chemical properties*, in *Advances in Synthesis Gas: Methods, Technologies and Applications.* 2023, Elsevier. p. 3–25.
93. Haghighatjoo, F. and M.R. Rahimpour, *Introduction to nonacidic impurities of natural gas: Particulates, condensates, mercury, nitrogen, helium*, in *Advances Natural Gas: Formation, Processing, and Applications. Volume 5: Natural Gas Impurities and Condensate Removal.* 2024, Elsevier. p. 3–24.
94. Udomsirichakorn, J., et al., *Effect of CaO on tar reforming to hydrogen-enriched gas with in-process CO2 capture in a bubbling fluidized bed biomass steam gasifier.* International Journal of Hydrogen Energy, 2013. 38(34): p. 14495–14504.
95. Zhang, K., et al., *Study on catalytic cracking of toluene as the model compound of tar by different catalysts.* Environmental Progress & Sustainable Energy, 2020. 39(1): p. 13295.
96. Yang, C.-L., et al., *Sulfur-anchoring synthesis of platinum intermetallic nanoparticle catalysts for fuel cells.* Science, 2021. 374(6566): p. 459–464.
97. Minteer, S., *Alcoholic Fuels.* 2016, CRC Press.
98. Baksi, S., et al., *Pre-treatment of lignocellulosic biomass: Review of various physico-chemical and biological methods influencing the extent of biomass depolymerization.* International Journal of Environmental Science and Technology, 2023. 20(12): p. 13895–13922.
99. Arruda Ferraz de Campos, V., et al., *A Waste-to-Energy technical approach: Syngas–biodiesel blend for power generation.* Energies, 2023. 16(21): p. 7384.
100. Devi, A., et al., *A perspective towards sustainable and economically viable approach of waste biorefineries through lignin valorization.* Biomass Conversion and Biorefinery, 2024: p. 1–25.
101. Adu, T.F., et al., *Decision support systems for waste-to-energy technologies: A systematic literature review of methods and future directions for sustainable implementation in Ghana.* Heliyon, 2025. **11**(3): p. e43353. https://doi.org/10.1016/j.heliyon.2025.e42353.
102. Sazali, N., et al., *New perspectives on fuel cell technology: A brief review.* Membranes, 2020. 10(5): p. 99.
103. Staffell, I., et al., *The role of hydrogen and fuel cells in the global energy system.* Energy & Environmental Science, 2019. 12(2): p. 463–491.
104. Wang, B., et al., *A review on gasification of municipal solid waste (MSW): Syngas production, tar formation, mineral transformation and industrial challenges.* International Journal of Hydrogen Energy, 2023. 48(69): p. 26676–26706.
105. Mishra, R., et al., *Co-gasification of solid waste and its impact on final product yields.* Journal of Cleaner Production, 2022. 374: p. 133989.
106. Rey, J., et al., *A review of cleaning technologies for biomass-derived syngas.* Fuel, 2024. 377: p. 132776.
107. Machado, M., et al., *Syngas enhancement for Fischer-Tropsch integration via solid oxide electrolyzer cell co-electrolysis with or without methane.* Iscience, 2024. **27**(10): p. 111014. https://doi.org/10.1016/j.isci.2024.111014.
108. Phillip, A., et al., *Enhanced biogas production using anaerobic co-digestion of animal waste and food waste: A review.* Journal of Scientific Research and Reports, 2024. 30(8): p. 761–781.

109. Shikhar, U., K. Hemmes, and T. Woudstra, *Exploring the possibility of using molten carbonate fuel cell for the flexible coproduction of hydrogen and power.* Frontiers in Energy Research, 2021. 9: p. 656490.
110. Antolini, E., *The stability of molten carbonate fuel cell electrodes: A review of recent improvements.* Applied Energy, 2011. 88(12): p. 4274–4293.
111. Abdelkareem, M.A., et al., *Environmental aspects of fuel cells: A review.* Science of The Total Environment, 2021. 752: p. 141803.
112. Lombardi, L., E. Carnevale, and A. Corti, *A review of technologies and performances of thermal treatment systems for energy recovery from waste.* Waste Management, 2015. 37: p. 26–44.
113. Sarker, T., et al., *Transformation of municipal solid waste to biofuel and biochemicals–A review.* International Journal of Environmental Science and Technology, 2025. 22(5): p. 3811–3832.
114. John, N., et al., *Physical conversion of biomass: Dewatering, drying, size reduction, densification, and separation*, in *Handbook of Biomass.* 2024, Springer. p. 987–1013.
115. Lin, Y., et al., *MSW characterization and preprocessing for biofuels and bioproducts*, in *Advances in Bioenergy.* 2023, Elsevier. p. 197–238.
116. Zafar, H., N. Peleato, and D. Roberts, *A review of the role of pre-treatment on the treatment of food waste using microbial fuel cells.* Environmental Technology Reviews, 2022. 11(1): p. 72–90.
117. Pen, M., J. Gomez, and J.G. Fierro, *New catalytic routes for syngas and hydrogen production.* Applied Catalysis A: General, 1996. 144(1–2): p. 7–57.
118. Imron, R., et al., *Producing hydrogen-rich syngas via microwave heating and cogasification: A systematic review.* Biofuel Research Journal, 2022. 9(1): p. 1573–1591.
119. Marcantonio, V., et al., *Main issues of the impact of tar, H2S, HCl and alkali metal from biomass-gasification derived syngas on the SOFC anode and the related gas cleaning technologies for feeding a SOFC system: A review.* International Journal of Hydrogen Energy, 2022. 47(1): p. 517–539.
120. Ausiello, A. and L. Micoli, *Treatment of Biogas for Feeding High Temperature Fuel Cells.* Springer, Cham.
121. Paglini, R., M. Gandiglio, and A. Lanzini, *Technologies for deep biogas purification and use in zero-emission fuel cells systems.* Energies, 2022. 15(10): p. 3551.
122. Medaiyese, F.J., et al., *From waste to energy: Enhancing fuel and hydrogen production through pyrolysis and in-line reforming of plastic wastes.* Sustainability, 2024. 16(12): p. 4973.
123. Cooke, S.J., *Industrial gases*, in *Handbook of Industrial Chemistry and Biotechnology.* 2017, Springer. p. 1301–1322.
124. Rusman, N. and M. Dahari, *A review on the current progress of metal hydrides material for solid-state hydrogen storage applications.* International Journal of Hydrogen Energy, 2016. 41(28): p. 12108–12126.
125. Lim, K.L., et al., *Solid-state materials and methods for hydrogen storage: A critical review.* Chemical Engineering & Technology: Industrial Chemistry-Plant Equipment-Process Engineering-Biotechnology, 2010. 33(2): p. 213–226.
126. Gvozdenac, D., et al., *High efficiency cogeneration: CHP and non-CHP energy.* Energy, 2017. 135: p. 269–278.
127. Wu, D. and R. Wang, *Combined cooling, heating and power: A review.* Progress in Energy and Combustion Science, 2006. 32(5–6): p. 459–495.
128. Ellamla, H.R., et al., *Current status of fuel cell based combined heat and power systems for residential sector.* Journal of Power Sources, 2015. 293: p. 312–328.
129. Mehrpooya, M., et al., *Technical performance analysis of a combined cooling heating and power (CCHP) system based on solid oxide fuel cell (SOFC) technology–A building application.* Energy Conversion and Management, 2019. 198: p. 111767.

130. Bajpai, P., *Pulp and Paper Industry: Energy Conservation*. 2016, Elsevier.
131. Wang, L., *Energy efficiency technologies for sustainable food processing*. Energy Efficiency, 2014. 7(5): p. 791–810.
132. Potter, N.N. and J.H. Hotchkiss, *Unit operations in food processing*, in *Food Science: Fifth Edition*. 1995, Springer. p. 69–89.
133. Bosmans, A., et al., *The crucial role of Waste-to-Energy technologies in enhanced landfill mining: A technology review*. Journal of Cleaner Production, 2013. 55: p. 10–23.
134. Demirbaş, A., *Biomass resource facilities and biomass conversion processing for fuels and chemicals*. Energy Conversion and Management, 2001. 42(11): p. 1357–1378.
135. Zhang, J., H. Cho, and A. Knizley, *Evaluation of financial incentives for combined heat and power (CHP) systems in US regions*. Renewable and Sustainable Energy Reviews, 2016. 59: p. 738–762.
136. Li, J., et al., *Incentive or constraint? Comprehensive impacts of green credit policy on industrial energy intensity*. Environmental Science and Pollution Research, 2023. 30(46): p. 103101–103118.
137. Khan, F., et al., *Enhancing biogas/biohydrogen utilization in dual-fuel engines using advanced machine learning algorithms*. International Journal of Hydrogen Energy, 2025. 112: p. 81–90.
138. Zhuang, L., et al., *Scalable microbial fuel cell (MFC) stack for continuous real wastewater treatment*. Bioresource Technology, 2012. 106: p. 82–88.
139. Zhao, J., et al., *Carbon emission reduction potential of municipal solid waste-to-energy under resource mismatch patterns: Evidence from urban agglomerations in China*. Resources, Conservation and Recycling, 2025. 217: p. 108188.
140. Yan, G., et al., *Artificial intelligence-powered study of a waste-to-energy system through optimization by regression-centered machine learning algorithms*. Energy, 2025. 320: p. 135142.
141. Bragatto, T., et al., *Optimization of an energy district for fuel cell electric vehicles: Cost scenarios of a real case study on a waste and recycling fleet*. International Journal of Hydrogen Energy, 2022. 47(95): p. 40156–40171.
142. Rahman, W., S. Yusup, and S.A. Mohammad. *Screening of fruit waste as substrate for microbial fuel cell (MFC)*. in *AIP Conference Proceedings*. 2021, AIP Publishing LLC.
143. Hibino, T., K. Kobayashi, and T. Hitomi, *Biomass solid oxide fuel cell using solid weed waste as fuel*. Electrochimica Acta, 2021. 388: p. 138681.
144. Usman, M., *Techno-economic and emissions comparison of waste-to-fuel via hydrothermal liquefaction, transesterification, and incineration*. Journal of Economy and Technology, 2025. 3: p. 237–250.
145. Lui, J., et al., *Life cycle assessment of waste-to-hydrogen systems for fuel cell electric buses in Glasgow, Scotland*. Bioresource Technology, 2022. 359: p. 127464.
146. Mann, S. *ExxonMobil, FuelCell Energy to Build CCS Pilot Plant*. 2023; Available from: https://www.wastetodaymagazine.com/news/exxon-mobil-fuelcell-energy-to-build-ccs-pilot-plant/?utm_source.
147. Connelly, E., *DOE's H2@Scale Initiative and Biogas Projects* 2019, U.S. Department of Energy, Office of Energy Efficiency & Renewable Energy.
148. U.S Department of Energy. *Project Selections for FOA 3082: Carbon Negative Shot Pilots*. (n.d.); Available from: https://www.energy.gov/hgeo/project-selections-foa-3082-carbon-negative-shot-pilots.
149. Lovells, H., *Australian Hydrogen Projects Paper*. 2020.
150. AfricanUnion. *Support for Waste-to-Energy Pilot Projects*. (n.d.); Available from: https://sdg7.africa-eu-energy-partnership.org/country-spotlight/waste-to-energy/.
151. Abifarin, J.K. and Y. Lu, *Statistics in enabling 2D materials: Optimization, predictive modelling, and data-driven discovery*. Materials Today Physics, 2025: p. 101814.

152. U.S. DepartmentofEnergy. *Comparison of Fuel Cell Technologies.* (n.d.); Available from: https://www.energy.gov/eere/fuelcells/comparison-fuel-cell-technologies?utm_source.
153. Paulillo, A., et al., *Decarbonising waste-to-energy: A life cycle assessment study.* Resources, Conservation and Recycling, 2024. 209: p. 107812.
154. Saito, H., *The role of iot in waste-to-energy technologies: Optimizing waste management.* American Journal of Internet of Things, 2025. 6(2): p. 13–189.
155. Kasulla, S., et al., *The role of Iot in waste management and sustainability.* Partners Universal International Research Journal, 2024. 3(2): p. 76–88.
156. Sosunova, I. and J. Porras, *IoT-enabled smart waste management systems for smart cities: A systematic review.* Ieee Access, 2022. 10: p. 73326–73363.
157. Alzahrani, A.I., S.H. Chauhdary, and A.A. Alshdadi, *Internet of Things (IoT)-based wastewater management in smart cities.* Electronics, 2023. 12(12): p. 2590.
158. Rane, N., S. Choudhary, and J. Rane, *Leading-edge Artificial Intelligence (AI), Machine Learning (ML), Blockchain, and Internet of Things (IoT) technologies for enhanced wastewater treatment systems.* Machine Learning (ML), Blockchain, and Internet of Things (IoT) Technologies for Enhanced Wastewater Treatment Systems (October 31, 2023), 2023.
159. Dahmani, S., *Computational intelligence for green cloud computing and digital waste management*, in *Computational Intelligence for Green Cloud Computing and Digital Waste Management.* 2024, IGI Global Scientific Publishing. p. 248–266.
160. Li, Y., et al., *Leveraging municipal solid waste management with plasma pyrolysis and iot: Strategies for energy byproducts and resource recovery.* Processes, 2025. 13(2): p. 321.
161. Kumar, P. and D. Choudhury, *Green technologies: Theories, systems, principles and applications*, in *Evaluating Environmental Processes and Technologies.* 2025, Springer. p. 149–195.
162. Adebowale, O.J. and E.A. Etukudoh, *Market assessment and strategic roadmap for fuel cell technology in stationary power applications.* Gyanshauryam International Scientific Refereed Research Journal, 2022. J 5(5): p. 175–210.
163. Akyüz, E., İ. Çobanoğlu, and B. Demircan, *Internet of things and cloud-based monitoring of an electrolysis and fuel cell system.* Journal of Scientific & Industrial Research (JSIR), 2025. 84(7): p. 731–741.
164. Sharma, P., et al., *An overview of artificial intelligence-based techniques for PEMFC system diagnosis.* IEEE Access, 2024. 12: p. 165708–165735.
165. Middha, R., N. Srivastava, and N. Saxena, *Chemical principles in waste segregation and recycling*, in *Waste Management for Smart Cities.* 2024, Springer. p. 1–46.
166. Omoseebi, A. and K.D. Jackson Tyler, *Energy efficiency: Smart grids and IoT-enabled energy management systems optimize electricity distribution, reduce waste, and integrate renewable energy sources.* 2023. https://www.researchgate.net/publication/389689947_Energy_Efficiency_Smart_grids_and_IoT-enabled_energy_management_systems_optimize_electricity_distribution_reduce_waste_and_integrate_renewable_energy_sources.
167. Zhang, X., G. Manogaran, and B. Muthu, *IoT enabled integrated system for green energy into smart cities.* Sustainable Energy Technologies and Assessments, 2021. 46: p. 101208.
168. Dave, D.M.K. and B.K. Mittapally, *Data integration and interoperability in IoT: Challenges, strategies and future direction. International Journal of Computer Engineering and Technology* (IJCET), 2024. 15: p. 45–60.
169. Albreem, M.A., et al., *Towards green Internet of Things (IoT) for a sustainable future in Gulf Cooperation Council countries: Current practices, challenges and future prospective.* Wireless Networks, 2023. 29(2): p. 539–567.

170. Khawaja, M.K., et al., *Waste-to-energy barriers and solutions for developing countries with limited water and energy resources.* Science of the Total Environment, 2024: 926: p. 172096.
171. Chowdury, M.S.K., et al., *Degradation mechanisms, long-term durability challenges, and mitigation methods for proton exchange membranes and membrane electrode assemblies with Pt/C electrocatalysts in low-temperature and high-temperature fuel cells: A comprehensive review.* Journal of Electroanalytical Chemistry, 2024. **975**: p. 118712.

6 Solar-Powered Greenhouse Farming

N R Govinthasamy, Kabilan C N, Kanishka S, Kiruthika E, and Rajavel M

6.1 INTRODUCTION

Agriculture has traditionally been the main source of human nutrition and the engine of world economic stability. As the world's population continues to grow and the need for food rises, the problem of guaranteeing food security has become more difficult. Farmers today face a complicated array of challenges, such as erratic weather patterns, dwindling arable land, severe water scarcity, and a shortage of agricultural workers, in addition to the need to increase crop production. These complex issues highlight the urgent need for a paradigm change away from traditional, resource-intensive farming practices, and towards sustainable, technologically advanced agricultural system (Table 6.1).

In recent years, a revolutionary method of upgrading agricultural operations has emerged through the merging of digital technology, particularly the Internet of

TABLE 6.1
List of Abbreviations

Abbreviation	Definition
IoT	Internet of Things
ESP32	Espressif Systems 32-bit Microcontroller Unit
MCU	Microcontroller Unit
DHT11	Digital Humidity and Temperature Sensor
PIR	Passive Infrared Sensor
LDR	Light Dependent Resistor
LCD	Liquid Crystal Display
LED	Light Emitting Diode
Wi-Fi	Wireless Fidelity
V	Volt
°C	Degree Celsius
%RH	Relative Humidity Percentage
mA	Milliampere
DC	Direct Current
PWM	Pulse Width Modulation
Blynk	IoT Mobile Platform for Device Control

 DOI: 10.1201/9781003671039-6

Things (IoT), with renewable energy solutions. Through intelligent, automated control of irrigation schedules, temperature regulation, and pest management, precision agriculture is empowered by IoT-driven smart farming, which makes it possible to continuously collect real-time environmental and crop-specific data. These technologies improve resource optimization and operational efficiency while reducing manual involvement.

However, reliance on traditional grid-based or fossil fuel-driven energy sources is a major barrier to the widespread adoption of IoT-based agricultural automation, especially in rural and off-grid areas. This reliance not only restricts deployment in isolated areas but also compromises these technologies' capacity for sustainability. The current study suggests a self-sufficient, solar-powered intelligent greenhouse farming system that may function without the need for external power infrastructure in order to overcome this difficulty. To provide adaptive and energy-efficient operational control, the system makes use of an ESP32 microcontroller coupled with a variety of environmental sensors, including motion detection modules, temperature sensors, and soil moisture sensors.

Through internet connectivity, the design enables smooth real-time oversight and management through Blynk-based remote monitoring and control. The suggested methodology greatly lowers operating costs, minimizes carbon emissions, and improves the resilience of agricultural operations in off-grid or energy-constrained areas by relying solely on solar photovoltaic (PV) electricity. In order to meet the productivity and environmental demands of the twenty-first century, this combination of IoT-driven automation and renewable energy autonomy offers a scalable and affordable route towards climate-resilient, resource-efficient, and sustainable agriculture.

6.2 LITERATURE SURVEY

6.2.1 IoT Architectures and Communication Platforms

The IoT has been used in recent smart agriculture developments to create scalable, reasonably priced greenhouse automation systems. Researchers have shown how real-time data collecting and control may be made possible using microcontrollers like the ESP32 and related wireless communication frameworks. Galon et al. (2025) created an ESP32-based solar greenhouse system with the Blynk application for real-time data visualization and actuator control, while Pandurangi and Teradal (2025) suggested a next-generation IoT-powered greenhouse model incorporating sophisticated sensing and control layers. Seetaram et al. (2024) and Godase et al. (2024) have employed comparable structures to efficiently control irrigation and climate variables. Bersani et al. (2022) investigated Industry 4.0-compatible IoT platforms appropriate for extensive greenhouse networks, while Zheng et al. (2023) presented a multi-sensor IoT framework that increased the accuracy of environmental monitoring. The integration of monitoring and control subsystems to improve crop productivity in controlled situations was further proven by Bhushanamu (2025).

6.2.2 Renewable Energy Integration and Energy Optimization

The incorporation of renewable energy, especially solar PV systems, has been well-documented as a sustainable way to power off-grid greenhouse activities, coinciding with the development of IoT technology. In their modelling and optimization of PV systems for rural agricultural settings, Liao et al. (2025) placed a strong emphasis on the design process's interpretability and transparency. An adaptive solar-powered greenhouse system that dynamically modified environmental control loads to correspond with the availability of solar energy in real time was demonstrated by Fukuyo et al. (2024). In order to balance light transmission and electricity generation, Ravishankar et al. (2022) investigated the integration of organic solar cells in greenhouse roofing. Meanwhile, Mu et al. (2022) suggested phase change material-based thermal storage for controlling temperature at night. Improved energy reliability has been demonstrated by hybrid renewable models that combine wind and solar energy to reduce intermittency problems. The viability of combining solar energy with automated environmental control in smaller-scale facilities was validated by early research by Aakanksha et al. (2021).

6.2.3 Automation, Control, and Machine Learning in Agriculture

High-precision agriculture with less manual intervention is made possible by the growing use of automation and artificial intelligence (AI) in greenhouse systems. SolarGrowNet, an autonomous IoT control network designed for Gerbera cultivation and able to manage several climate parameters at once, was presented by Aher et al. (2025). While Ndunagu et al. (2022) created a wireless soil moisture-based irrigation network that reduced human labour while optimizing production, Tace et al. (2022) integrated IoT and machine learning to optimize irrigation schedules, boosting water use efficiency. Despite early research showing farmer reluctance to use autonomous systems due to worries about safety, dependability, and responsibility, later studies have emphasized the significance of regulatory frameworks and demonstrated operational resilience to promote adoption. Particularly in arid and water-stressed areas, the integration of AI and IoT has produced quantifiable improvements in resource efficiency, yield forecasting precision, and resilience against climatic unpredictability. Utilizing soil data and prediction algorithms, irrigation innovations have also reduced waste and guaranteed ideal crop hydration.

6.2.4 Reviews, Surveys, and Data-Driven Insights

The development of smart greenhouse systems has been charted by numerous thorough surveys, which have also highlighted current research needs. In their comprehensive assessment of greenhouse monitoring solutions, Abou-mehdi-Hassani et al. (2025) highlighted the difficulties with standardization and interoperability. In their assessment of IoT-enabled sustainable agriculture techniques, Kumar et al. (2024) emphasized the collaboration between real-time environmental sensors and AI analytics. Dandrifosse et al. (2024) enhanced the dependability of weather-dependent decision-making by automatically regulating the quality of climatic information, while Bettucci et al. (2024) illustrated the function of agricultural data analytics in optimizing the use of machinery. When taken as a whole, these studies provide a solid basis for investigating next-generation solar-powered intelligent greenhouse systems, especially those that seek to integrate scalable IoT architectures, AI-driven decision-making, and energy autonomy for high-yield, sustainable precision agriculture.

Agriculture underpins global food security but faces challenges from unpredictable weather, water scarcity, and diminishing arable land. Conventional farming methods are labour-intensive and inefficient in resource utilization. The integration of IoT with renewable energy provides a sustainable path forward.

This study proposes a solar-powered greenhouse system that autonomously manages irrigation, temperature, lighting, and pest control. Built around an ESP32 NodeMCU controller, the design enables smart decision-making and remote accessibility through the Blynk IoT platform.

6.3 EXISTING SYSTEM

Greenhouse systems are essential for increasing crop output in the fields of smart agriculture and controlled environment farming because they offer a buffer against erratic weather patterns. Over time, greenhouse designs have changed from being straightforward passive structures to complex, technologically advanced spaces that can control vital elements, including soil moisture levels, temperature, humidity, and light intensity.

Many traditional greenhouse systems still mostly rely on manual labour, energy sources derived from fossil fuels, and non-adaptive control mechanisms in spite of these developments. These restrictions frequently lead to variable crop performance, less sustainable power management overall, and less-than-ideal energy use. Combining solar energy with automated greenhouse control has become a viable strategy as the world moves towards sustainable agriculture and the use of clean energy. A unified, self-sustaining, and adaptable greenhouse framework must be developed because most current systems show a lack of smooth coordination between intelligent environmental regulation and solar energy management.

Conventional farming methods rely heavily on labour-intensive and inefficient manual monitoring and control of irrigation, pest control, and other crucial activities. Despite the introduction of certain automated greenhouse systems, these systems typically use localized controllers and connected power supplies without features for remote monitoring or sustainable energy use. These systems' main drawbacks are as follows:

1. **Excessive Water Consumption**: Inefficient irrigation methods contribute to significant water wastage.
2. **High Labour Requirements**: Continuous manual inspection of soil moisture, plant health, and pest activity increases operational workload.
3. **Dependence on Conventional Power Sources**: Reliance on grid electricity restricts deployment in remote or off-grid locations.
4. **Absence of Real-Time Alerts**: Lack of immediate notifications delays corrective actions and reduces system responsiveness.

An intelligent, IoT-enabled, solar-powered greenhouse system that can monitor the climate in real time, adjust its control, and run sustainably is needed to overcome these constraints. A system like this can greatly increase resource efficiency, lessen the need for human involvement, and encourage ecologically friendly farming methods.

6.4 PROPOSED SYSTEM

A solar-powered, IoT-enabled greenhouse automation framework designed to offer an intelligent, energy-efficient, and sustainable growing environment is introduced by the suggested system. The system, which is powered by a 6 V PV panel, is designed to operate as environmentally friendly as possible, especially in off-grid or isolated agricultural environments. To continuously monitor soil moisture, ambient temperature, humidity, and light intensity in real time, the ESP32 NodeMCU microcontroller acts as the central processing unit and interfaces with environmental sensors.

6.4.1 Important Features and Functional Elements Include

Automated Irrigation Control: Water resources are conserved by ensuring irrigation only happens when required, which is controlled by soil moisture sensors.

Mitigation and Pest Detection: When a passive infrared (PIR) motion sensor detects pest activity, automatic countermeasures like targeted sprays or auditory deterrents are triggered.

Renewable Power Management: Without depending on traditional power sources, a specialized solar-based energy management plan guarantees operational dependability.

IoT-Based Remote Access: Through online or mobile applications, farmers may monitor in real time, receive alert notifications, and exercise control thanks to a smooth interaction with the Blynk IoT platform.

Self-Regulating Closed-Loop Control: This automated design keeps the ideal environmental conditions for plant growth while reducing the need for human intervention.

Cost-Efficiency and Scalability: Designed for minimal operating costs and modular expansion, this technology can be adopted in a variety of agricultural scenarios.

This self-sufficient, integrated system offers a data-driven, sustainable method of greenhouse management that boosts output while guaranteeing resource preservation and environmental responsibility (Figure 6.1).

6.4.2 Block Diagram

The Smart Solar-Powered Greenhouse Farming System controls the environmental factors necessary for the best possible plant growth using an automation architecture based on microcontrollers and fully powered by renewable solar energy. Solar

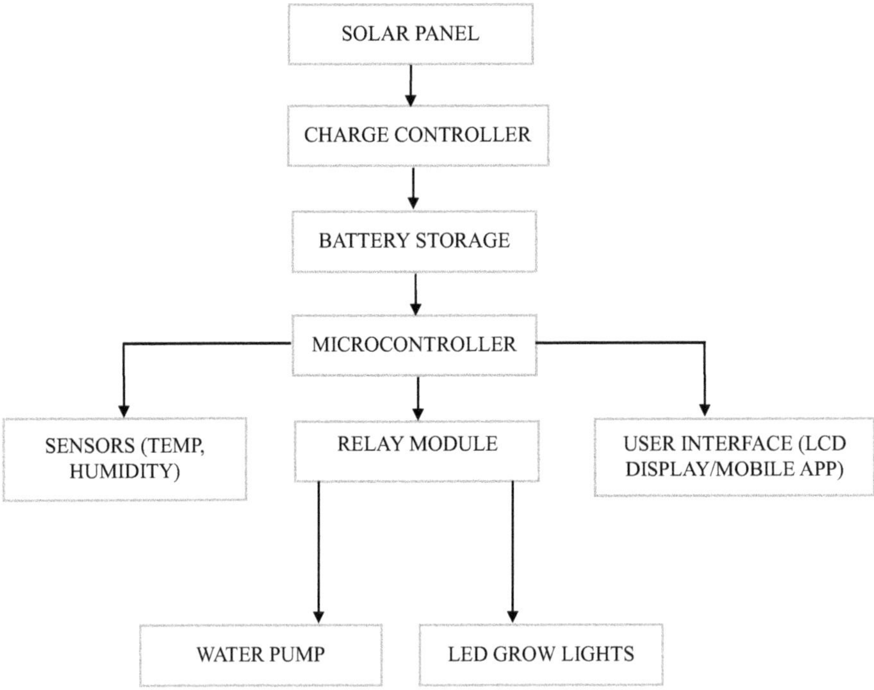

FIGURE 6.1 Block diagram of Smart Solar-Powered Greenhouse Farming system.

radiation is captured by a PV module and sent to a battery storage unit via a charge controller, guaranteeing continuous functioning throughout the day and night.

At the heart of the system is an ESP8266 microcontroller, which collects and analyses data in real time from a variety of environmental sensors that measure the soil moisture, temperature, and relative humidity in the greenhouse microclimate. Using a relay module and pre-programmed threshold logic, the microcontroller controls linked devices. For example, it turns on the water pump when soil moisture drops below a predetermined level and turns on LED grow lights when ambient light levels are too low to support photosynthetic activity.

The system includes both a mobile-based interface made possible by the Blynk IoT platform and an LCD display enabling on-site status feedback for both local and remote human–machine interaction. The ESP8266's integrated Wi-Fi connectivity enables real-time monitoring, remote control, and alarm notifications from any place with internet access by sending live sensor readings and actuator status to the Blynk application.

Arduino-based firmware on the ESP8266 implements the operational logic of the system, which includes sensor calibration, decision-making algorithms, and actuator scheduling. A completely autonomous, energy-efficient, and remotely accessible greenhouse management system is the outcome of the smooth integration of renewable energy harvesting, environmental monitoring, automated actuation, and IoT-enabled user interaction. By eliminating resource waste, lowering manual labour, and facilitating precision agriculture on a large scale, the suggested design promotes sustainable farming methods.

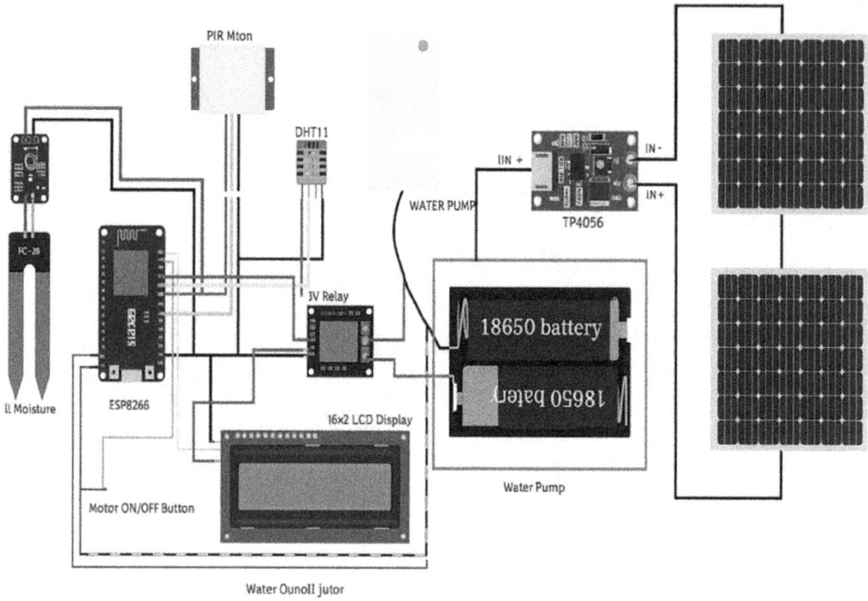

FIGURE 6.2 Circuit diagram of Smart Solar-Powered Greenhouse Farming system.

6.4.3 Circuit Diagram

The circuit diagram in Figure 6.2 illustrates the electrical configuration of the solar-powered greenhouse farming system, detailing the interconnection of each component to ensure efficient operation and long-term sustainability. The solar panel serves as the primary energy source, harvesting renewable power that is regulated by a charge controller to prevent battery overcharging or deep discharging, thereby extending battery lifespan and ensuring consistent performance.

At the core of the system, the microcontroller functions as the central processing unit, acquiring real-time data from environmental sensors and executing control algorithms for optimal operation. High-power devices, including water pumps and LED grow lights, are operated via a relay module, which provides electrical isolation to enhance safety and reliability.

The system incorporates dual user interfaces: a local LCD display for on-site monitoring and a mobile application for remote access, offering flexibility and redundancy in system control. Its modular architecture supports scalability, enabling seamless integration of future enhancements such as advanced climate regulation or automated nutrient management. Overall, the proposed circuit design delivers a robust, energy-efficient, and self-sustaining solution for greenhouse operations, requiring minimal user intervention.

The following formulas are used and proposed for better hardware results.

6.4.3.1 Incident Solar Power on PV Panel

$$\mathrm{Pin}(t) = G(t) \cdot \mathrm{Apv} \cdot \cos(\theta) \qquad \text{Eqn. (1)}$$

G(t): solar irradiance at time t (W/m^2)
Apv: effective PV panel area (m^2)
θ: angle of incidence between sunlight and panel

6.4.3.2 Output Power from PV Module

$$Pou(t) = \eta pv \cdot Pin(t) - Ploss \quad \text{Eqn. (2)}$$

ηpv: efficiency of solar cell
Ploss: losses (due to wiring, dust, shading, etc.)

6.4.3.3 Battery Energy Balance

$$SOC(t+1) = SOC(t) + \left(\left(\eta_c \cdot P_{ch}(t)\right) - \left(P_{dis}(t) / \eta_d / C_{batt}\right)\right)_{.\Delta t} \quad \text{Eqn. (3)}$$

SOC(t): battery charge at time t
$P_{ch}(t)$: charging power (W)
$P_{dis}(t)$: discharging power (W)
$\eta_{c,\eta d}$: charging/discharging efficiency
C_{batt}: battery capacity (Wh)

6.4.3.4 Crop Growth & Yield

Light-Use Efficiency (LUE) Model:

$$Y - \varepsilon \cdot \int APAR(t)\, dt \quad \text{Eqn. (4)}$$

Y: biomass or yield (kg)
ε: crop-specific light-use efficiency (g/MJ)
APAR(t): absorbed photosynthetically active radiation (MJ/m^2)
T: growing season length

6.4.3.5 Water Use Efficiency (WUE)

$$WUE = Y / ETc \quad \text{Eqn. (5)}$$

Y: yield (kg)
ETc: crop evapotranspiration (mm or L/m^2)

6.4.3.6 IoT-Based Smart Control

$$u(t) = f\left(Tin(t), RH(t), SM(t), I(t)\right) \quad \text{Eqn. (6)}$$

u(t): control action (pump ON/OFF, fan speed, etc.)
SM: soil moisture (%)
I: incident solar radiation (W/m^2)
Function f(·): Decision-making logic from AI/IoT system

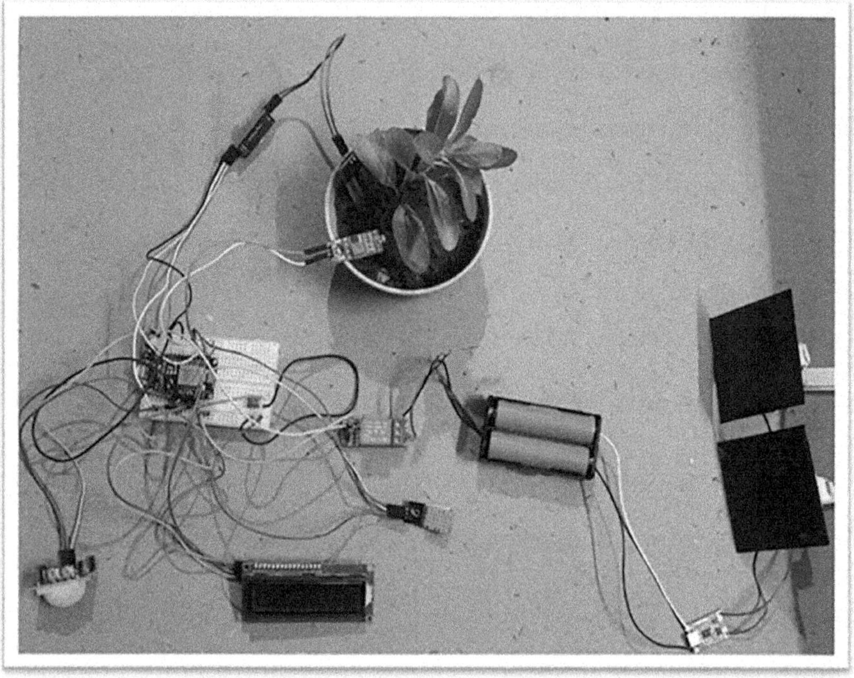

FIGURE 6.3 Smart Solar-Powered Greenhouse Farming system setup.

The performance of the solar-powered intelligent greenhouse system was quantitatively assessed using the developed equations. Calculations were made for battery sizing, load requirements, and daily solar energy yield to guarantee continuous operation in a range of environmental circumstances.

In order to maintain ideal temperature regulation, thermal balancing and ventilation needs were examined, and models of humidity and evapotranspiration enabled accurate irrigation scheduling. These models were used to determine the potential biomass production and crop water use efficiency, indicating the system's ability to use resources sustainably. The efficiency of the suggested design for data-driven, high-yield precision agriculture was confirmed by the experimental graphs, which also supported the theoretical expectations (Figures 6.3–6.5).

6.4.4 Hardware Setup and Results

The central processing unit is a microcontroller, which communicates with a variety of environmental sensors to gather data on temperature, humidity, soil moisture, and light intensity in real time. The processing of this sensor data allows parts like LED grow lights and water pumps to be activated automatically. A relay module is used to control actuators, providing electrical isolation and shielding low-voltage control circuitry from high-voltage power components.

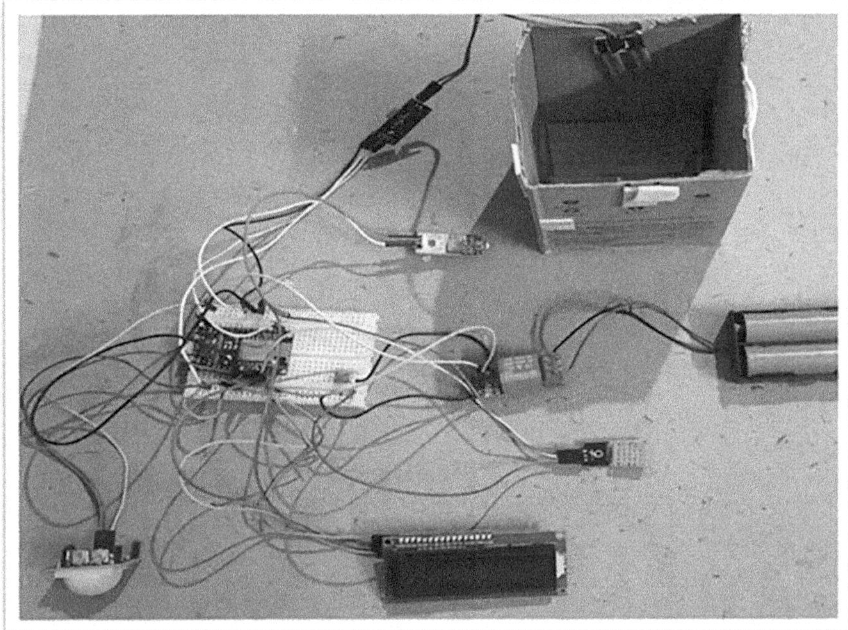

FIGURE 6.4 Smart Solar-Powered Greenhouse Farming system in different ambient setups.

The system can operate in two modes: remotely via a mobile application and locally through an LCD display for on-site monitoring. System accessibility and user flexibility are improved by this hybrid interaction model. Future improvements, such as automated nutrient dosing systems or precision climate regulation devices, can be seamlessly integrated because of the modular design. A stable and expandable framework for self-sufficient greenhouse management with little assistance from humans is offered by the hardware architecture.

6.4.5 IoT Integration

The cloud-based framework for real-time monitoring and management of the solar-powered smart greenhouse is provided by the Blynk IoT platform, which integrates sensors, actuators, and the ESP8266 microcontroller with an adaptable mobile and online interface. This dashboard allows users to remotely manage elements like irrigation pumps and exhaust fans as well as view real-time environmental parameters like temperature, humidity, light intensity, and soil wetness. With the help of the ESP8266's Wi-Fi module, continuous cloud connectivity is made possible, guaranteeing worldwide accessibility and enabling automated reactions to sensor inputs to maintain ideal growing conditions without the need for user intervention (Tables 6.2 and 6.3).

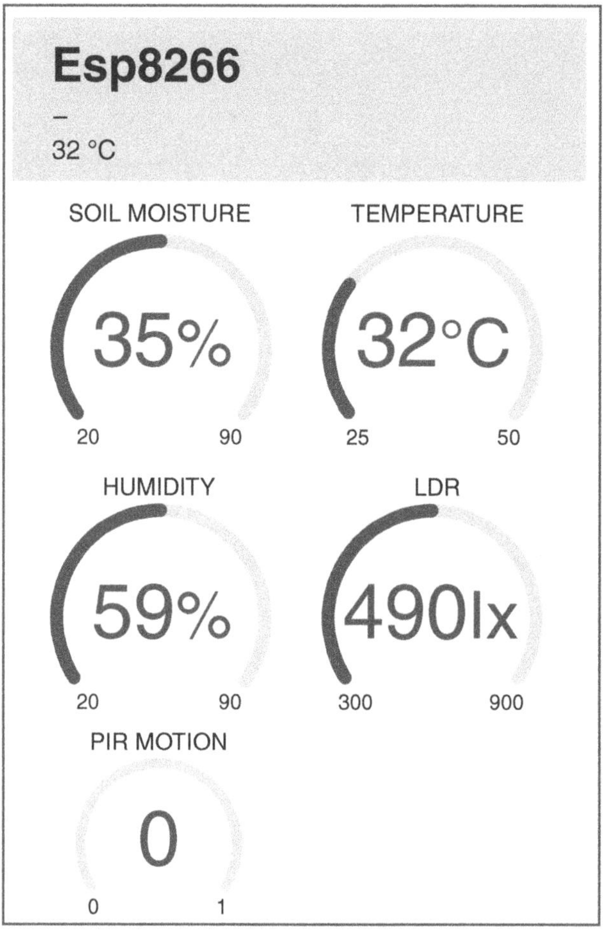

FIGURE 6.5 IoT integration with Blynk IoT platform.

From Figure 6.6 and Table 6.4, comparative analysis impact the better solution for sensor-based automation that enabled the environmental control system to efficiently maintain ideal greenhouse conditions. The DHT11 sensor, which turned on the exhaust fan when temperatures exceeded 32°C and turned it off when they reached about 28°C, allowed the temperature to be controlled from its peak of 36–28°C. The DHT11 sensor, which started recording relative humidity at 48%, stabilized it at 65%. When humidity dropped below 60%, it activated misting fans, and when it reached 70%, it stopped working.

A soil moisture sensor that started the irrigation pump at values below 35% and halted it at about 65% was used to maintain the soil moisture, which had fallen below 30%, between 60% and 65%. An LDR sensor kept the light intensity, which was measured at less than 450 lux during low sunshine, between 500 lux and 700 lux. It activated LED grow lights whenever the lighting fell below the 450 lux threshold.

TABLE 6.2
Comparative parameter analysis of proposed system

Parameter	Natural/Uncontrolled Condition	Controlled Condition (Proposed System)	Method of Achievement
Temperature (°C)	36°C (peak)	28°C maintained	DHT11 sensor triggers exhaust fan above 32°C until ~28°C is reached
Humidity (%)	48%	65% stabilized	DHT11 sensor activates misting fans when humidity <60% and stops at ~70%
Soil Moisture (%)	Dropped below 30%	Maintained between 60% and 65%	Soil moisture sensor controls irrigation pump ON at <35% and OFF at ~65%
Light Intensity (lux)	< 450 lux during low sunlight	500–700 lux maintained	LDR sensor switches on LED grow lights when light <450 lux

TABLE 6.3
Sensor specifications

Sensor	Measured Parameter	Range/Specification	Purpose in System
DHT11	Temperature & Humidity	0–50°C, 20–90%RH	Controls fans and misting for climate regulation
Soil Moisture Sensor	Soil water content	0–100% analog output	Automates irrigation cycle based on thresholds
PIR Sensor	Motion/Pest activity	Up to 6 m detection range	Triggers alarm or deterrent mechanism
LDR Sensor	Ambient light intensity	0–1000 lux (approx.)	Controls LED grow lights
Solar Panel	Electrical power	6 V, 2 W output	Primary renewable energy source

The comparison analysis in Figure 6.7 emphasizes how much better solar-powered intelligent greenhouse systems are than conventional methods. The automated setup achieves 95% system efficiency, which is almost twice as high as traditional techniques. Compared to 50% in conventional systems, water utilization is optimized at 70% of daily needs, guaranteeing greater sustainability.

Automation maintains healthier growing conditions by regulating soil moisture more consistently. Furthermore, whereas technical involvement accounts for 70% of the effort, automation significantly reduces the need for manual labour from 70% to 30%. The intelligent system is a future-ready solution for smart agriculture since it exhibits increased production, resource efficiency, and decreased reliance on human involvement.

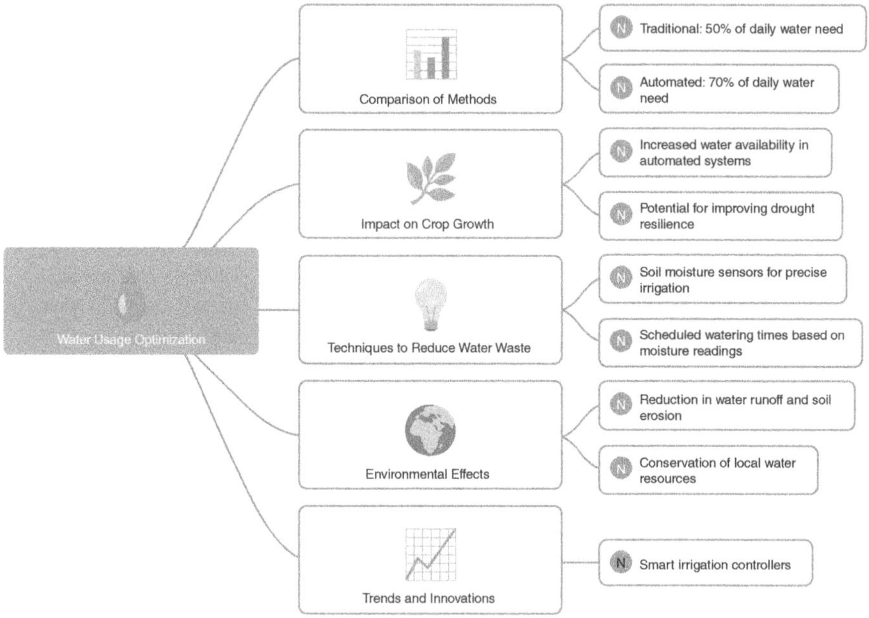

FIGURE 6.6 Water usage optimization with greenhouse farming.

TABLE 6.4
Comparative performance of automated and traditional greenhouse system

Parameter	Traditional Method	Automated System	Improvement (%)
System Efficiency	50%	95%	+45%
Water Usage (of daily req.)	50%	70%	+20%
Labour Requirement	65%	25%	−40%
Soil Moisture (optimal range)	50–55%	60–65%	+10%
Temperature (°C)	36	28	−8
Humidity (% RH)	55	65	+10
Crop Yield	Baseline	+15–20%	+15–20%

6.5 CONCLUSION

A feasible route to resource-efficient and sustainable agriculture is provided by the combination of solar-powered greenhouse farming and intelligent automation. The current design reduces reliance on traditional power sources by integrating renewable energy, surpassing the 79% operational efficiency of the prior automation-based planting method. Real-time environmental sensing, which includes temperature, light intensity, soil moisture, and human presence, along with automated control systems, has made it possible to precisely regulate ventilation and irrigation, which has improved crop health, significantly reduced water consumption, and stabilized the microclimate.

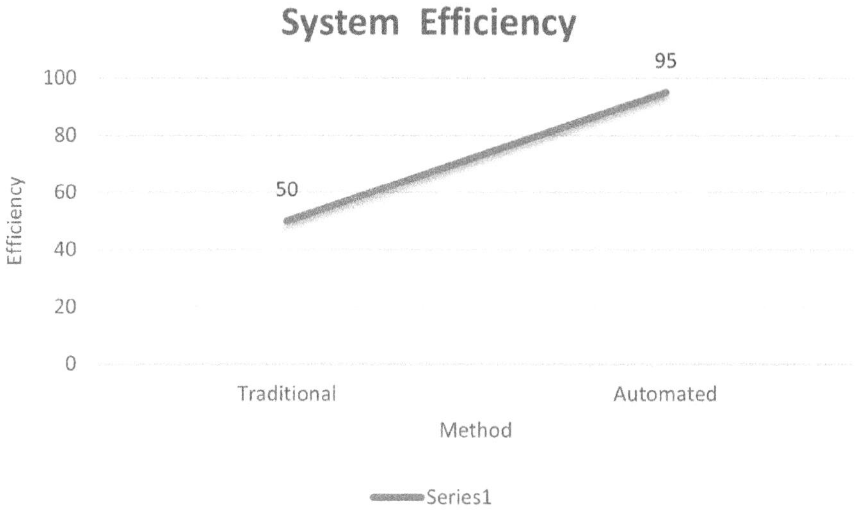

FIGURE 6.7A–D Comparative analysis of traditional versus automated greenhouse systems.

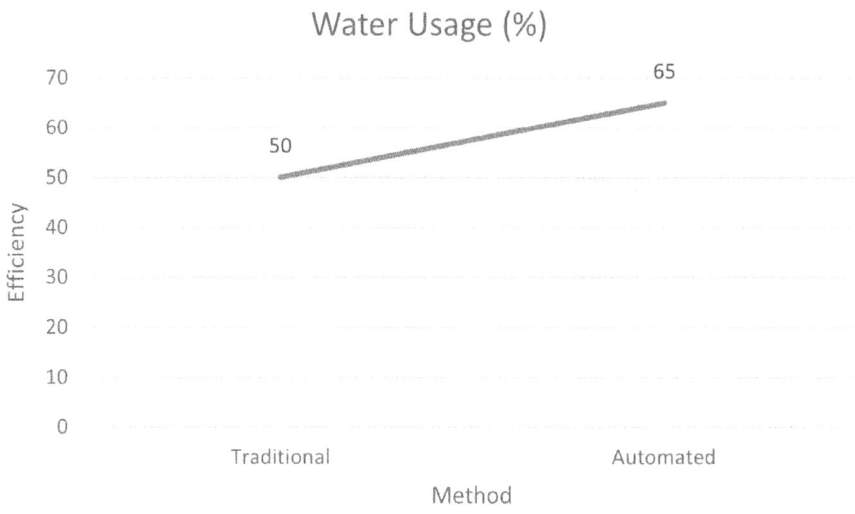

FIGURE 6.7A–D (Continued)

Significant energy savings of about 1.5 kWh per day, or more than 500 kWh yearly, are achieved by the system's total energy autonomy during the hottest parts of the day, aided by battery backup. From Figure 6.7, experimental results of comparative parameter analysis of proposed system with traditional method show both economic and environmental benefits, with an 86% resource optimization efficiency and up to 30% water use reduction. Additionally, it may be used in both urban and rural settings due to its scalable and modular design, which guarantees flexibility across different crop varieties and greenhouse dimensions.

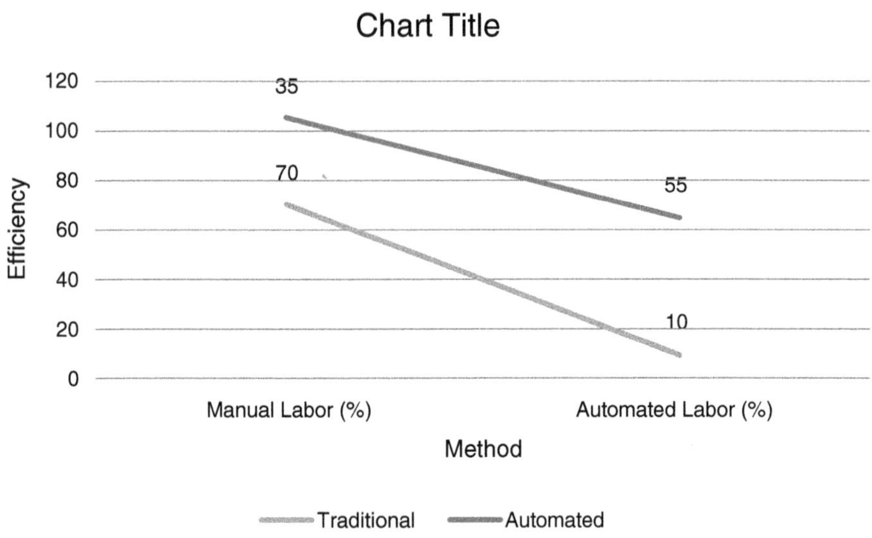

FIGURE 6.7A–D (Continued)

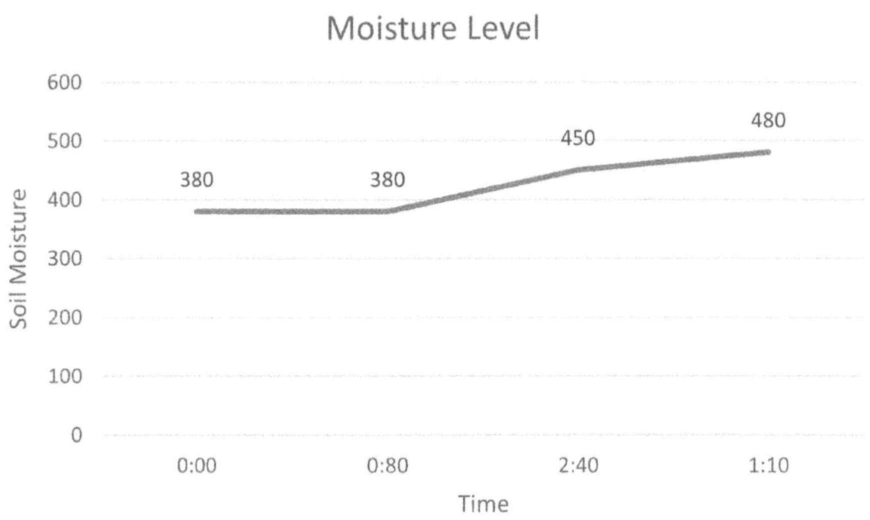

FIGURE 6.7A–D (Continued)

Predictive maintenance, early disease identification, and predictive environmental control are some of the ways that the integration of AI and machine learning is anticipated to improve system performance going forward. With these developments, this strategy could become the standard for smart, renewable-energy-powered greenhouse systems in contemporary farming.

REFERENCES

A. Aakanksha, C. Anand, S. Shameem Akhter, and S. Nikitha. "Smart Greenhouse Automation Using Solar Power." *International Journal of Creative Research Thoughts (IJCRT)*, Vol. 9, No. 6, June 2021.

F. Abou-mehdi-Hassani, A. Zaguia, H. A. Bouh, and A. Mkhida. "Systematic Literature Review of Smart Greenhouse Monitoring." *SN Computer Science*, Vol. 6, No. 95, January 2025.

V. N. Aher, V. More, and R. S. Pol. "SolarGrowNet: Autonomous Greenhouse Monitoring and Control System for Gerbera." *Journal of Information Systems Engineering and Management*, Vol. 10, e-ISSN: 2468-4376, pp. 239–254, March 2025.

C. Bersani, C. Ruggiero, R. Sacile, A. Soussi, and E. Zero. "Internet of Things Approaches for Monitoring and Control of Smart Greenhouses in Industry 4.0." *Energies*, Vol. 15, Article 3834, May 2022. https://doi.org/10.3390/en15103834.

F. Bettucci, M. Sozzi, M. Benetti, and L. Sartori. "A Data-Driven Approach to Agricultural Machinery Working States Analysis during Ploughing Operations." *Smart Agricultural Technology (Elsevier)*, Vol. 8, July 2024. Article ID: 100511.

M. B. N. Bhushanamu. "IoT Based Greenhouse Monitoring and Control System." *International Journal of Scientific Research in Engineering and Management*, Vol. 9, pp. 1–9, April 2025.

H. J. F. Fukuyo, R. A. B. Ramos, and A. C. P. Canonicato. "Solar-Powered, Internet-of-Things-Based Smart Greenhouse with Adaptive Environment Control for Sustainable Agriculture Nursery." *IEEE Photovoltaic Specialists Conference (PVSC)*, Seattle, WA, USA, pp. 0526–0532. 2024, doi: 10.1109/PVSC57443.2024.10749686

S. Dandrifosse, D. Rosillon, and A. Jago. "Automatic Quality Control of Weather Data for Timely Decisions in Agriculture." *Smart Agricultural Technology (Elsevier)*, Vol. 8, April 2024. Paper ID: 100445.

M. L. Q. Galon, M. V. R. Tumaliwan, and M. M. Sejera. "Automated Monitoring and Control System of Solar Greenhouse Using ESP32 and Blynk Application." *Engineering Proceedings*, Vol. 92, Article 57, 2025.

V. Godase, Y. Jadhav, and K. Vishal. "IoT Based Greenhouse Monitoring and Controlling System." *International Journal for Scientific Research & Development*, Vol. 12, No. 3, pp. 138–140, 2024. ISSN (online): 2321–0613.

A. Khedkar, B. Bhise, and D. Shinde. "Solar Powered Greenhouse Monitoring Using IoT." *Journal of Science and Technology*, Vol. 6, Special Issue 1, August 2021.

V. Kumar, K. V. Sharma, N. Kedam, and A. Patel. "A Comprehensive Review on Smart and Sustainable Agriculture Using IoT Technologies." *Smart Agricultural Technology (Elsevier)*, Vol. 8, June 2024. Article Id: 100487.

R. Liao, M. Manfren, and B. Nastasi. "Off-grid PV Systems Modelling and Optimisation for Rural Communities—Leveraging Understandability and Interpretability of Modelling Tools." *Energy*, Vol. 324, pp. 1–16, June 2025.

M. Mu, S. Zhang, S. Yang, and Y. Wang. "Phase Change Materials Applied in Agricultural Greenhouses." *Journal of Energy Storage*, Vol. 49, Article 104100, May 2022.

J. N. Ndunagu, et al. "Development of a Wireless Sensor Network and IoT-Based Smart Irrigation System." *Applied and Environmental Soil Science*, Vol. 2022, Issue 1, Article ID 7678570, 2022.

B. Pandurangi and S. G. Teradal. "Next-Generation IoT-Powered Smart Greenhouse." *International Research Journal of Modernization in Engineering Technology and Science*, Vol. 7, No. 6, p. 1708–1716, June 2025.

E. Ravishankar, R. E. Booth, J. A. Hollingsworth, H. Ade, H. Sederoff, J. F. DeCarolis, and B. T. O'Connor. "Organic Solar-Powered Greenhouse Performance Optimization and Global Economic Opportunity." *Energy & Environmental Science*, Vol. 15, p. 1659, 2022.

J. Seetaram, A. Bhavya, C. Tarun, and V. Sameera. "Internet of Things (IoT) Based Greenhouse Monitoring and Controlling System Using ESP-32." *International Journal of Advanced Research in Computer and Communication Engineering*, Vol. 13, No. 6, June 2024.

Y. Tace, et al. "Smart Irrigation System Based on IoT and Machine Learning." *Energy Reports*, Vol. 8, pp. 1025–1036, 2022.

B. Zheng, S. Tian, C. Rao, Q. Wang, C. Liang and B. Tan, "Wireless Smart Greenhouse Management System Based on Multi-sensor of IoT," *2023 IEEE 3rd International Conference on Electronic Communications, Internet of Things and Big Data (ICEIB)*, Taichung, Taiwan, pp. 95–99, 2023. doi: 10.1109/ICEIB57887.2023.10170112.

7 IoT-Based Monitoring and Maintenance Solution for Solar Street Lighting System and Model Optimization for Enhanced Solar Efficiency in Dust-Prone Environments

Abhishek Kharwar, Debabrata Bej, and Nikhil Sadashiv Dani

7.1 INTRODUCTION

Street lighting systems are highly necessary to provide safety, security, and usability in public areas at night. In urban areas, they consume 15–40% of the total electricity demand, while globally it is responsible for approximately 3% of total energy usage.[1,2] In earlier times the street lighting systems were predominantly grid-based. Grid-based electricity is derived from fossilized energy. About 67% of world's electricity generation is from fossilized energy.[3] This is responsible for the harmful emissions and climate change. In recent times the solar street lighting systems are on the playing field of street lighting systems because they are powered by solar energy, and the use of solar-based street lighting systems has become favored as they can work independently from the electrical grid. This also renders them suitable for remote, rural, or disaster zones where grid access is not available or dependable. Off-grid lighting provides uninterrupted service even during electricity outages. This lighting system is also eco-friendly, as it reduces carbon emissions and fossil fuel use considerably.[4] It also supports climate change mitigation and sustainable development goals.[4]

The electrical performance of photovoltaic (PV) modules largely depends on the intensity and quality of incident solar irradiance.[5] Photovoltaic cells exhibit typical

DOI: 10.1201/9781003671039-7

current–voltage (I–V) and power–voltage (P–V) characteristics that differ with irradiance levels. Higher solar insolation on the panel surface is caused by greater photon absorption, increasing power generation and hence the efficiency of the PV system.[6,7] However, real-world conditions of operation introduce a variety of factors that are responsible for the degradation in solar panel efficiency. Such losses can be caused by several sources, that is, optical losses because of reflection and spectral mismatch, thermal losses because of elevated cell temperatures, electrical losses in the system (e.g., resistive wiring losses and inefficiencies of inverter), and environmental losses because of shading, snow, and especially deposition of dust. Excess energy losses can vary between 5% and 30% of the overall daily watt-hour load based on the system configuration, the angle of installation, and local climatic conditions.[8]

Among all environmental factors, dust deposition is one of the most important causes of efficiency loss in PV systems. Dust accumulation on the PV module surface, particularly in dry and semi-arid zones like the Sahel, can lead to considerable power output decreases—up to 78% per year in the most severe cases.[9] Under less severe conditions, efficiency may be reduced by as much as 15% by soil, dirt, and atmospheric particulate deposits on the modules' glass surface.[10] Dust impacts PV panel performance via direct deposition—creating a physical obstruction that scatters or absorbs light—and environmental dispersal processes, wherein wind-borne particulates unevenly settle and produce non-uniform shading and hotspots. The cumulative accumulation nature of dust, especially in areas with low rainfall or where poor maintenance is implemented on arbitrarily long schedules, makes it a chronic and growing factor in the analysis and optimization of the performance of the PV system.

In recent days, the emergence of Internet of Things (IoT) technologies with solar lighting infrastructure has created numerous possibilities for remote monitoring and smart control. An IoT-based solar and piezoelectric hybrid system to conserve energy, fault detection, and dimming using vehicle/ambient sensors was presented in.[11] But this system mainly concerns energy conservation, power consumption, and fault detection but excludes soiling effects—a big reason behind efficiency loss in dusty environments. In a research in Ref. [12], the authors modeled soiling and temperature-based efficiency loss, but these are still empirical without implementations to automated, real-time systems. In Ref. [13. 14], authors developed image processing technology for the detection of dust deposition on solar panels. In Ref. [15], researchers employed optical imaging of PV surfaces and aerosol measurements to acquire PV panel's dust accumulation and aerosols on PV performance. Simulation-driven maintenance frameworks with real-time data, predictive models, and controlled alerting are however absent in this work. In Ref. [16], scientists designed a robotic cleaning technique for PV panels to forecast and schedule through machine learning for minimization of soil effects, particularly concentrated in ground-mounted PV systems in arid and semi-arid regions. In Ref. [17], the authors came up with an early warning solar panel disruption due to dust accumulation using the Support Vector Machine (SVM) algorithm. In Ref. [18], authors came up with an intelligent, AI-powered dust detection system employing camera and convolutional neural network (CNN) on a Raspberry Pi for solar panel cleanliness classification in real time, and predictive maintenance development. However, to the authors' knowledge, relatively little research exists on simulations of merging cloud-collected sensor measurements with extraterrestrial radiation modeling or numerical methods (e.g., Levenberg-Marquardt) to quantify the loss of efficiency due to dust under varying environmental conditions.

In this chapter, authors propose an IoT-based solution for monitoring and maintaining solar street lights, focusing on the impact of dust accumulation on panel efficiency. The system measures voltage and current data, processes it through a microcontroller, and uses cloud-based analysis to estimate efficiency and generate maintenance alerts. We incorporate Simulink simulations to estimate maintenance alerts and employ the Levenberg-Marquardt algorithm to predict extraterrestrial radiation, introducing correction factors for sunshine hours and temperature models. Our analysis identifies a 2-micron dust layer as the critical maintenance threshold, with the temperature-based model proving economically optimal.

This chapter is structured as follows. Methodology of the study is depicted in Section 7.2. Section 7.3 describes the simulation study using Simulink and the experimental study for monitoring and maintenance of the solar street lights. Section 7.4 describes the results and discussions of the study. This work is eventually concluded in Section 7.5.

7.2 METHODOLOGY

We present the proposed IoT-based solution for the monitoring and maintenance of solar street lights in Figure 7.1, utilizing a halogen-based solar setup for experimentation. Dust accumulation significantly impacts solar panel output, so our approach specifically accounts for its effect on panel efficiency. The solar panel generates DC current, with a voltage sensor measuring the panel's output voltage and a current

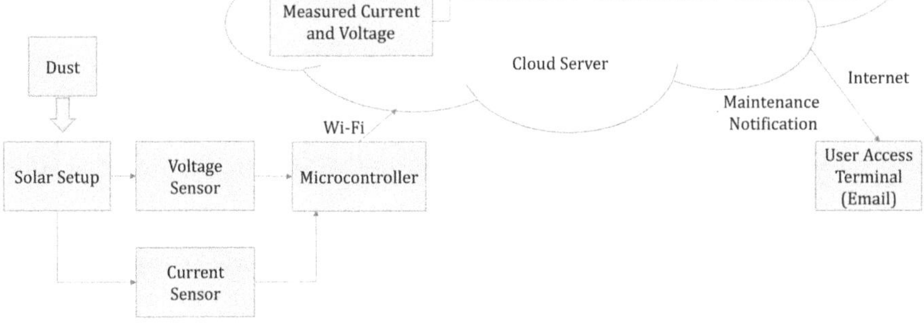

FIGURE 7.1 Proposed IoT-based Solution for Monitoring and Maintenance of Solar Street Lights.

transducer capturing the output current. A microcontroller processes the sensor data and transmits it to a cloud server via Wi-Fi. In the cloud server, the irradiance measured by a solar energy meter, along with the current and voltage data, is used to estimate the panel's efficiency. This estimated efficiency is then compared to the standard efficiency, calculated using the open-circuit voltage (V_{OC}) and short-circuit current (I_{SC}). Based on this comparison, a maintenance notification is generated, which can be sent to the user for further action.

In the context, an investigation has been undertaken to evaluate diverse parameters associated with solar radiation estimation by using simulation and theoretical study. The estimation process is facilitated by the utilization of distinct models, specifically temperature and maximum sunshine hour models. Critical to this investigation is the characterization of extraterrestrial radiation, which is contingent upon certain foundational variables, namely the solar constant, latitude, hour angle, and declination angle. The latitude under scrutiny is 10.80°N, thereby situating the investigation within the northern hemisphere. The hour angle is a pivotal metric for this study. Furthermore, this angle is intricately linked with the declination angle, denoting the angular deviation of the sun from Earth's equatorial plane. The interplay between these angles, coupled with the day of the month, bears significance in the pursuit of precise radiation estimates. Notably, terrestrial radiation, a parameter of paramount importance, hinges on empirical coefficients, maximum sunshine hour data, and average day length, which will be expounded upon in forthcoming sections of this study. Specifically, the inclination or tilt angle plays a pivotal role in determining terrestrial radiation levels. Moreover, this study acknowledges the presence of dust, uniformly distributed over the surface of solar panels, as a factor influencing the radiation dynamics.

7.3 SIMULATION STUDY AND EXPERIMENTAL STUDY

7.3.1 Simulation Study

A simulation study was carried out to assess maintenance considerations while considering the influence of ambient temperature and irradiance values. It was observed that maintenance alerts are triggered when the operational efficiency of the system falls below established standards. The simulation utilized solar panels, batteries, and a MATLAB program for analysis.

In the month of March, the irradiance value on the tilted surface was computed to be 277 Wm^{-2} using an empirical relation derived from the maximum sunshine hour model. Simultaneously, the ambient room temperature was recorded at 29°C. The calculation of tilt radiation considered several parameters, as detailed in Table 7.1.

The irradiance value of March is estimated as 301 Wm^{-2} on the tilted surface from the empirical relation by the temperature model, and the room temperature was measured as 29°C. The following are the parameters that were used to estimate the tilt radiation observed in Table 7.1.

Based on this, information is collected by the Simulink in the form of "0" or "1" about maintenance. Maintenance is required when the live efficiency is less than

TABLE 7.1
Parameters for estimation of radiation for maximum sunshine hour model and temperature model

Parameters	Value of Parameters for Estimation of Radiation for Maximum Sunshine Hour Model	Value of Parameters for Estimation of Radiation for Temperature Model
R^{-}_{g} (MJ/m2)	23.47	26.08
R^{-}_{d} (MJ/m2)	6.46	6.42
$R^{-}q$	1.031	1.0316
$R^{-}v$	0.99113	0.991133
ρ_g	0.2	0.2

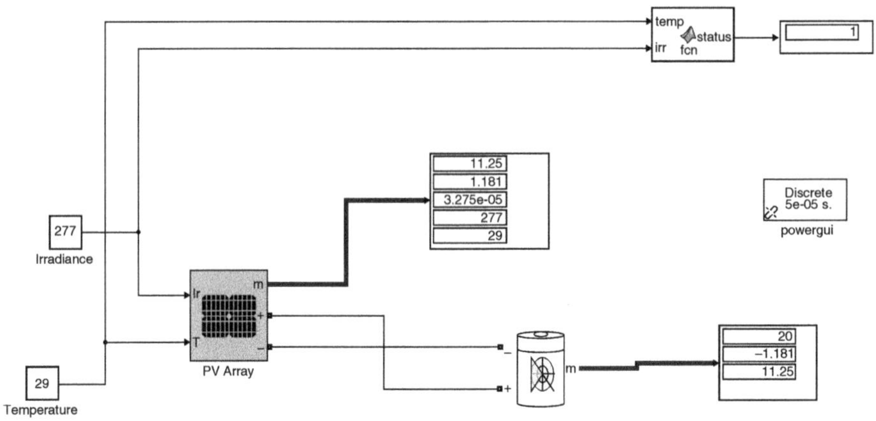

FIGURE 7.2 Simulink schematic of the PV system assessing maintenance alerts under varying ambient temperature and irradiance conditions.

the standard efficiency. A value of “0” indicates that no maintenance is required, whereas a value of “1” implies maintenance is necessary. A schematic of the simulation of maintenance is shown in Figure 7.2.

In this section, the algorithm is applied to gather maintenance-related data by varying the input parameters associated with solar irradiance and temperature. Observations are drawn from Figure 7.3. Furthermore, the maintenance messages depicted in the aforementioned figure can be discerned and analyzed for insights.

7.3.2 Experimental Study

In this section, we elaborate on the experimental setup, as depicted in Figure 7.4. The primary objective is to optimize solar radiation exposure by positioning the setup perpendicular to the sun’s rays at a normal tilt angle. To ensure optimal performance

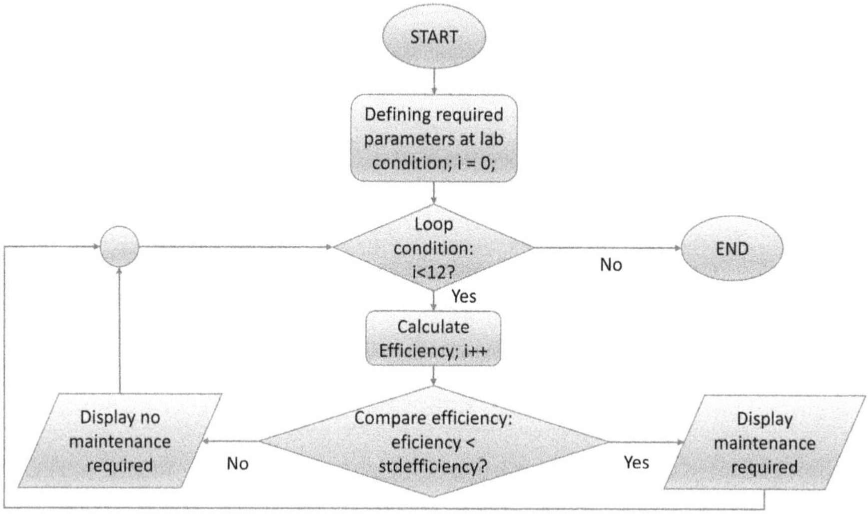

FIGURE 7.3 Flowchart of the proposed maintenance system for solar street lighting

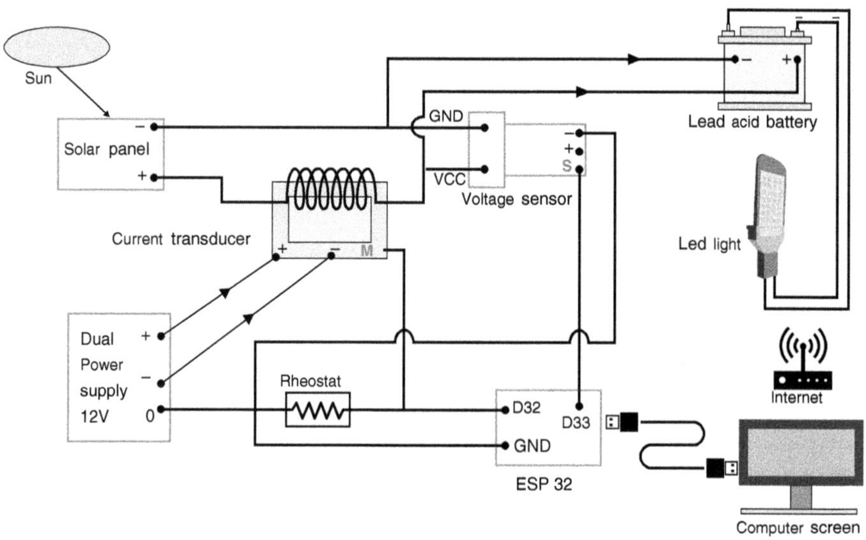

FIGURE 7.4 Schematic diagram of experimental setup

during different seasons, the arrays are adjusted to specific angles relative to the latitude. For efficient winter operation, the array angle equals the latitude plus 15°. While for effective summer operation, the array angle equals the latitude minus 15°. This alignment strategy maintains the solar panel surface's perpendicularity to the sun's rays throughout the year, with a specific tilt angle set at 10.80° N[22]. The positive terminal of the solar panel is connected via a wire to a current transducer with seven turns, utilizing the Hall effect for current measurement. The transducer operates on a 12 V DC supply, sourced from a dual power supply. Current measurement occurs through the transducer's M terminal; however, the transducer operates on a 12 V DC supply. It necessitates a connected resistance of 75.4 Ω. One end of this resistance is linked to the M terminal, while the other connects to 0 V from the dual power supply. Subsequently, this resistance terminal interfaces with the ESP32 microcontroller, specifically at pin D32. Moreover, the positive terminal of the panel links to the VCC terminal of a voltage sensor module, which incorporates a voltage divider for voltage level adjustment. This is crucial because the ESP32 is limited to handling no more than 3.3 V. The voltage sensor module employs a voltage divider comprising two resistors in series, namely 30 K and 7.5 K, to ensure the microcontroller receives a suitable voltage level. The panel's negative terminal is attached to the GND terminal of the voltage sensor module. Simultaneously, the S terminal of the sensor connects to pin D33 of the ESP32, while the sensor's negative terminal ties into the microcontroller's GND pin. Furthermore, the ESP32 establishes a connection with the 'thing-speak' IoT platform via a laptop. Consequently, LED lighting derives power from the battery to meet its illumination requirements. The experiment comprises two distinct scenarios: the first scenario involves the absence of dust, exemplified by Figure 7.5(a). In contrast, the second scenario, represented in Figure 7.5(b), introduces varying layers of dust ranging from 0 to 7 μm. The determination of dust thickness relies on area density, which involves the ratio of the dust particle's volume to the panel's surface area. Consequently, the dust is evenly distributed across the panel's surface. The study aims to explore the correlation between solar panel efficiency and dust thickness. Cement particles are employed as dust particles, characterized by specific properties: density (ρ_{cement} = 3120 kg/m^3), volume (58.653 cm^3), and panel area (0.52 m^2).

In Figure 7.8, a comprehensive block diagram illustrating the prediction process of extraterrestrial irradiance is presented. The process begins with the acquisition of input and output data derived from empirical estimations, which form the foundation for the subsequent analysis. This data is fed into the Levenberg–Marquardt Algorithm (LMA). The LMA is selected here over other techniques as it efficiently combines the fast convergence of the Gauss–Newton method with the stability of gradient descent, which makes it particularly effective for optimizing nonlinear models. This is well-suited for training neural networks and other nonlinear systems. The LMA algorithm initiates by processing the input data, where it iteratively adjusts the parameters to minimize the difference between the predicted and actual values. This phase is critical as it refines the model to accurately represent the underlying physical phenomena. Following the algorithmic adjustments, the dataset undergoes three distinct phases: training, validation, and testing. During the training phase, the model learns from the input data, iteratively improving its predictive capabilities. The validation phase then evaluates the model's performance on a separate dataset, ensuring that it generalizes well to unseen data. Finally, the testing phase rigorously assesses the model's accuracy and reliability. Once the model has been successfully trained, validated, and tested, a MATLAB function is generated. This function encapsulates the predictive model, making it readily deployable for further analysis or integration into larger systems. The final step involves using this MATLAB function to predict extraterrestrial irradiance. By applying the model to new input data, the function generates accurate predictions of extraterrestrial irradiance, providing valuable insights for applications such as solar energy research and atmospheric studies.

7.4 RESULTS AND DISCUSSION

7.4.1 Monitoring of Solar Panel

A simulation study was conducted to evaluate the need for maintenance of solar panels under varying radiation conditions, employing both temperature and maximum sunshine hour models. The results indicated that, in most cases, solar panel efficiency exceeded 7%, with occasional instances of efficiency falling below this threshold when there was no dust present. These efficiency levels closely aligned with the standard efficiency, which was determined based on parameters such as open-circuit voltage, short-circuit current, and fill factor (Figure 7.5). The study integrated an automated monitoring system that collected maintenance-related messages. These messages were displayed on a computer interface (see Figure 7.6 (a)) and also transmitted to email (see Figure 7.6 (b)) through IoT platforms like ThingSpeak and If This Then That (IFTTT) whenever the actual efficiency dipped below the predefined standard efficiency thresholds. Under the temperature model, maximum efficiency was attained at approximately 280 Wm^{-2} and reached a minimum of around 210 Wm^{-2}, as discussed in the section on the influence of temperature on electrical parameters. In contrast, the maximum sunshine hour model showed maximum efficiency at 250 Wm^{-2}, declining to a minimum of 225 Wm^{-2}. These observations were made under dust-free conditions. The standard efficiency, derived from the defined parameters, stood at 12.45%. In the presence of dust, the thickness of the dust layer on the solar panel influenced efficiency, as exemplified in upcoming section. Dust layer thickness was determined by area density, and the findings clearly indicated that efficiency decreased in dusty conditions compared to dust-free scenarios.

7.4.2 Effect of Dust Layer on Efficiency

The impact of dust accumulation on solar panel efficiency is of paramount importance, as it is evident that a significant reduction occurs when the panel is contaminated with dust. The thickness of the dust layer plays a crucial role in this context.

For solar panels constructed from polycrystalline materials, the typical efficiency ranges between 10% and 13%. However, the presence of dust leads to efficiency degradation, potentially resulting in a substantial cumulative loss in power generation. Regular maintenance, involving the cleaning of the panels, can mitigate this issue. Therefore, obtaining timely maintenance information is imperative. In the case of the temperature-based model, observations from Figure 7.7 (a) suggest that as irradiance magnitude increases, efficiency decreases up to an irradiance level of 240 Wm^{-2}. Beyond this point, a transient increase in efficiency occurs, reaching approximately 10% under dust-free conditions before declining again for a dust thickness of 0 μm. However, regardless of dust layer thickness

FIGURE 7.5 Experimental setup: (a) without dust and (b) with dust.

(2, 5, and 7 μm), efficiency exhibits a consistent decline with increasing irradiance. Notably, efficiency experiences an upturn above an irradiance level of 280 Wm^{-2}, peaking at around 7.5%, 5%, and 5% for dust layer thicknesses of 2, 5, and 7 μm, respectively, at specific irradiance magnitudes. Conversely, the lowest efficiencies were observed at approximately 3.5%, 2.5%, and 2.5% for dust layer thicknesses of 2, 5, and 7 μm, respectively, also at specific irradiance levels. Conversely, in the case of the sunshine-based model, the findings from Figure 7.7 (b) indicate that as irradiance magnitude increases, efficiency initially rises to nearly 10% and

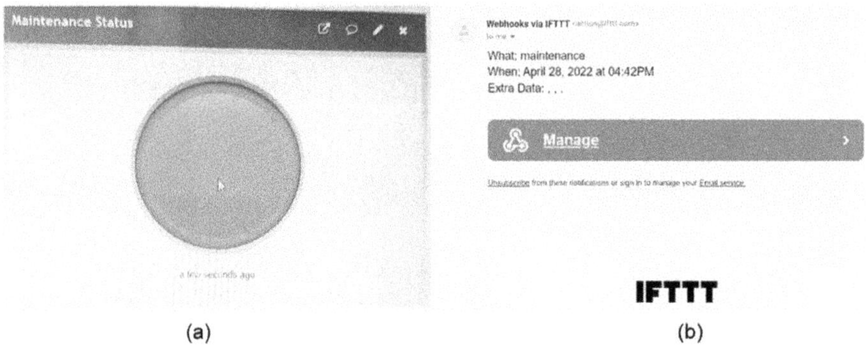

FIGURE 7.6 Maintenance alert: (a) on computer screen and (b) in an email.

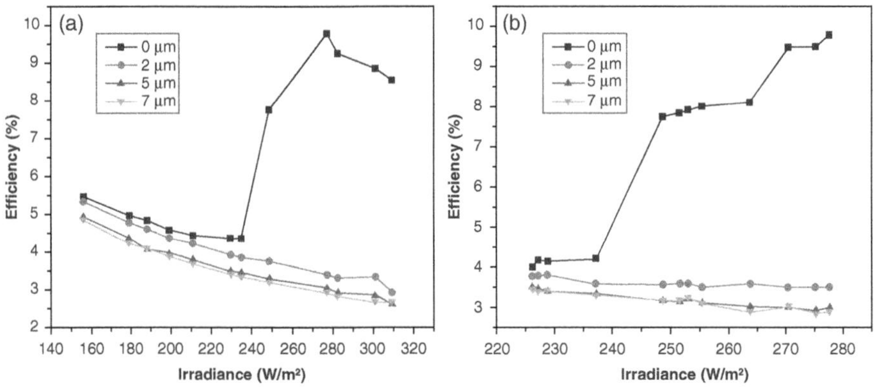

FIGURE 7.7 Effect of dust layer thickness on efficiency: (a) for temperature-based model and (b) for maximum sunshine hour model.

then decreases as irradiance continues to increase, particularly in the absence of dust. Similar to the temperature model, efficiency shows a consistent decline with increasing irradiance regardless of dust layer thickness 2, 5, and 7 µm. However, there is an efficiency increase above an irradiance level of 270 Wm^{-2} for a 2 µm dust layer. The highest efficiencies were approximately 7%, 3.5%, and 3.5% for dust layer thicknesses of 2, 5, and 7 µm, respectively, whereas the lowest efficiencies were approximately 3.5%, 3%, and 3% for the same respective dust layer thicknesses. Overall, the data from both models, as presented in Figures 7.7(a) and 7.7(b), reveal negligible decreases in efficiency beyond a 2 µm dust layer, irrespective of dust thicknesses of 5 and 7 µm. This trend underscores the importance of addressing the two primary issues affecting efficiency: the heating effect and the propensity of dust to adhere to the solar panel. Therefore, it is recommended that maintenance be conducted once a dust layer thickness reaches 2 µm (Figure 7.8).

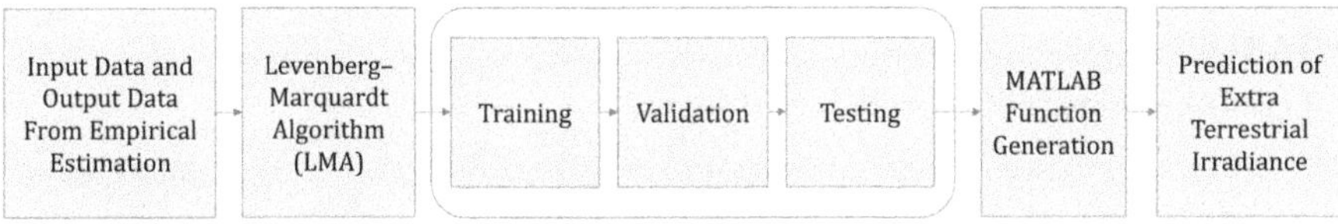

FIGURE 7.8 Block diagram overview of the prediction of extraterrestrial irradiance.

7.4.3 Prediction of Solar Radiation Through ANN

The following image is an example of predicting the global irradiance value through ANN. Parameters are selected, and finally, results from ANN came for the global solar irradiance value for the tilted surface. For example, input parameters Month-1; global radiation at the outer surface of the atmosphere is 30 MJ/m2/day, empirical constant, a = 0.5 and b = 0.3; the maximum sunshine hour is 10 hours; day length is 13 hours then output global radiation on the horizontal surface of earth is 20.2737 Wm^{-2} d.

7.4.4 Economic Feasibility

This section undertakes an economic feasibility analysis using acquired experimental data. It is observed that an increase in the thickness of the dust layer leads to a proportional rise in power losses for both the temperature and sunshine-based models. These losses are quantified over the span of one year. Notably, the temperature model exhibits superior performance compared to the sunshine model in mitigating power losses, resulting in overall lower losses. This comparison underscores the temperature model's enhanced effectiveness at the specific location under consideration. A comprehensive analysis of the provided table reaffirms this observation. It highlights that the minimum economic losses are associated with the temperature model and a dust layer thickness of 2 μm, outperforming all other combinations of models and thicknesses. Consequently, based on the findings of this study, it is strongly recommended that maintenance interventions for streetlight solar panels be initiated when the dust layer thickness reaches 2 μm. This proactive approach can significantly minimize economic losses, optimizing resource allocation and reducing unnecessary expenditures. Economic feasibility for different layers is shown in Table 7.2.

7.4.5 Correction Factor for Solar Radiation

In this section, the correction factor is determined based on actual solar radiation measurements for both the temperature and maximum sunshine hour models. The precise measurement of the daily average global solar radiation value for the month of April is established at 310 Wm^{-2} d^{-1}. Specifically, for the temperature model, this value is slightly lower at 290 Wm^{-2} d^{-1}, while the maximum sunshine hour model yields a measurement of 276.26 Wm^{-2} d^{-1} for April. Correction factor is the ratio of actual to estimated radiation. Furthermore, it is proposed that, in the case of the

TABLE 7.2
Economic feasibility for different layers

Sl. No.	Dust Layer (μm)	Power Loss in Temperature Model (kWh)	Power Loss in Maximum Sunshine Model (kWh)	Domestic Tariff per Unit in Kerala (INR)	Total Cost for Temperature Model (INR)	Total Cost for Maximum Sunshine Model (INR)
1	2	12.19	15.29	6.4	78.0	97.86
2	5	13.75	16.92	6.4	88.0	108.29
3	7	14.05	17.14	6.4	89.92	109.7

temperature model, the correction factor stands at 1.0689, whereas for the maximum sunshine hour model, the correction factor is determined to be 1.122. It is evident that the correction factor is lower for the temperature model compared to the sunshine hour model.

7.5 CONCLUSION

A simulation and an experimental study are devoted to investigating real-world scenarios, illustrating the experimental setup under conditions involving both dust and dust-free environment. Data are collected via IoT platforms such as ThingSpeak, and notifications regarding maintenance status are conveyed through computer screens and email, facilitated by IFTTT platform. The presence of dust significantly reduces the efficiency of solar panel. Additionally, correction factor has been provided for temperature and maximum sunshine hour model, that is, 1.0689 and 1.122, respectively. The error stands for temperature and maximum sunshine hour model at 6.89% and 12.2%, respectively. Moreover, at 2-micron, total power loss is 12.19 and 15.29 kWh for the temperature and maximum sunshine hour models. Therefore, the temperature-based model emerges as the more efficient choice for the current location, with significantly lower error rates compared to the maximum sunshine hour model. The findings, supported by real-world data collected via IoT platforms, offer valuable insights for optimizing solar panel performance in dusty environments.

REFERENCES

1 Belloni, E., L. Lunghi, I. Celli, A. Massaccesi, C. Moscatiello, & L. Martirano, Applications of hybrid solar streetlamps: electrical performance measurements and development of algorithms for their optimal management, *IEEE Access,* 12 (2024) 119602–119614.
2 Em-Udom, J., & N. Jaisumroum, SDT smart hybrid streetlight pole design utilizing renewable energy for a smart city in Thailand, *Smart Grids and Sustainable Energy,* 8(3) (2023) 16.
3 US Energy Information Administration. *Independent statistics & analysis, petroleum & other liquids.* (2013). https://www.nrc.gov/docs/ML1325/ML13253A313.pdf.

4 Nagaraja, M. R., W. K. Biswas, & C. P. Selvan, Advancements and challenges in solar photovoltaic technologies: Enhancing technical performance for sustainable clean energy–a review, *Solar Energy Advances*, 5 (2024) 100084.
5 Tian, W., Y. Wang, J. Ren, & L. Zhu, Effect of urban climate on building integrated photovoltaics performance, *Energy Conversion and Management*, 48(1) (2007) 1–8.
6 Rekioua, D., & E. Matagne, *Optimization of photovoltaic power systems: Modelization, simulation and control*, Springer Science & Business Media, 2012.
7 Mekhilef, S., R. Saidur, & M. Kamalisarvestani, Effect of dust, humidity and air velocity on efficiency of photovoltaic cells, *Renewable and Sustainable Energy Reviews*, 16(5) (2012) 2920–2925.
8 Panguloori, R., P. Mishra, & S. Kumar, Power distribution architectures to improve system efficiency of centralized medium scale PV street lighting system, *Solar Energy*, 97 (2013) 405–413.
9 Ndiaye, A., C. M. Ke´be´, P. A. Ndiaye, A. Charki, A. Kobi, & V. Sambou, Impact of dust on the photovoltaic (PV) modules characteristics after an exposition year in Sahelian environment: The case of Senegal, *International Journal of Physical Sciences*, 8 (21) (2013) 1166–1173.
10 Kaldellis, J., & A. Kokala, Quantifying the decrease of the photovoltaic panels' energy yield due to phenomena of natural air pollution disposal, *Energy*, 35 (12) (2010) 4862–4869.
11 Kazmi, S. N. A., A. Ulasyar, & M. F. N. Khan, IoT based energy efficient smart street lighting technique with air quality monitoring. In *2020 14th International Conference on Open Source Systems and Technologies (ICOSST)* (2020, December) (pp. 1–6). IEEE.
12 Mani, M., & R. Pillai, Impact of dust on solar photovoltaic (PV) performance: Research status, challenges and recommendations, *Renewable and Sustainable Energy Reviews*, 14(9) (2010) 3124–3131.
13 Dantas, G. M., O. L. C. Mendes, S. M. Maia, & A. R. de Alexandria, Dust detection in solar panel using image processing techniques: A review. *Research, Society and Development*, 9(8) (2020) e321985107–e321985107.
14 Yfantis, E., & A. Fayed, A camera system for detecting dust and other deposits on solar panels, *Advances in Image and Video Processing*, 2(5) (2014) 1–10.
15 Roumpakias, E., & T. Stamatelos, Surface dust and aerosol effects on the performance of grid-connected photovoltaic systems, *Sustainability*, 12(2) (2020) 569.
16 Al-Humairi, A., E. Khalis, Z. A. Al Hemyari, & P. Jung, Machine learning-based predictive maintenance for photovoltaic systems, *AI*, 6(7) (2025) 133.
17 Mukasir, H., F. A. Samman, A. Achmad, & M. Itasari, Predictive maintenance system using support vector machine algorithm for dust cleaning on solar panels. In *2023 International Conference on Modeling & E-Information Research, Artificial Learning and Digital Applications (ICMERALDA)* (2023, November) (pp. 50–55). IEEE.
18 Elkhoundafi, O., A. Elmrabet, & R. Elgouri, Design and manufacturing of an intelligent dust detector for solar panels using artificial intelligence, *Ecological Engineering & Environmental Technology*, 26(7) (2025) 216–226.

8 IoT Innovations for Marine Renewable Energies

Legal, Technical, and Policy Pathways for Decentralized Ocean Energy Systems

Sumanta Bhattacharya and Bhavneet Kaur Sachdev

8.1 INTRODUCTION

Marine renewable energy (MRE) has been identified as a potential source of energy to meet the global transition towards low-carbon energy regimes. The world's oceans, which span more than 70% of the Earth's surface, are a vast source of renewable energy. Various forms of marine energy, such as tidal, wave, offshore wind, ocean thermal energy conversion (OTEC), and salinity gradient energy, can be utilized. Each source has specific advantages and technological challenges. Tidal energy depends upon the predictable gravitational relationship between the Earth, moon, and sun, and it is one of the most stable sources of renewable energy (Lewis et al., 2011). Wave motion is a direct form of energy concentrated on ocean surfaces, powered by surface winds and oceanic swells, and is vast yet highly fluctuating, demanding robust conversion technologies (Falcão, 2010). The offshore wind energy sector has experienced the most rapid growth, the result of improvements in materials and technologies for wind turbines and floating support systems, it being the most commercially developed MRE sector (IRENA, 2021). OTEC, which uses temperature differences between surface seawater (warm) and deep seawater (cold), and saline gradient energy using differences in osmotic pressure between salt and fresh water, are at earlier stages of development but have potential in long-term generation of renewable energy (Khan et al., 2009).

Despite their potential, there are considerable challenges associated with the exploitation of MRE technologies and the offshore environment in general. Severe marine conditions bring challenges in operation and maintenance, and variable resource availability poses issues in grid control integration (Esteban & Leary, 2012). Advanced digital technologies are essential in dealing with these challenges, to vary the design, monitor, and manage the performance. In offshore energy systems, digitalization is no longer an additional tool but a fundamental enabler of efficiency,

 DOI: 10.1201/9781003671039-8

cost-competitiveness, and system robustness. Sensors capture these signals continuously as the analysis is integrated into operating methods, digitalized with other analytics and communicated from all wind turbines to wind parks and regional operations centres. For instance, digital twins – computerized counterparts to physical objects – have grown in importance in offshore energy, with installations – which simulate environmental loads and predict maintenance regimes (Tao et al., 2019). Those types of innovations may be necessary to cut downtime, prolong infrastructure life, and enhance safety in locations where manual inspections are dangerous and expensive.

There is a necessity for digitalization within marine energy as a result of the worldwide transition to smart and integrated energy systems. As an example, the extension of offshore wind farms on scales of tens or hundreds of megawatts would require advanced integration on shore. Digital application permits energy production in real-time on the basis of the weather models, and thus enables smoother interconnection with the national power system (Hahn et al., 2020). In addition, digital technologies underpin adaptive control systems, which help to smooth out swings in power production from marine renewables by enabling storage and demand-side management. Without a digital transformation, marine renewable projects risk being uncompetitive against economic and unpredictable operational costs. Accordingly, digitalization is the key enabler of a future-proof offshore energy ecosystem, rendering it more sustainable, scalable, and robust to climate and market uncertainties.

This chapter explores the roles of different components of Internet of Things (IoT) in the development of MRE resources. It also explains the scope of technological innovations in the area of MRE resources. Thereafter, this chapter focuses on the legal, regulatory, and social considerations in the development of MRE resources. It also emphasizes the key factors that need to be considered for ensuring sustainable growth of ocean renewable energy. Then, the challenges in the deployment of technologically equipped MRE are addressed followed by some possible recommendations for the future.

8.2 POTENTIAL ROLE OF IOT IN TRANSFORMING MARINE ENERGY

The IoT is the great transforming enabler for MRE. IoT is described as an extensive array of smart devices that have sensors and embedded communication modules, which are capable of gathering, transmitting, or processing data on the cloud in real-time (Atzori et al., 2010). For the marine energy sector, IoT has the potential to transform all segments of the value chain, beginning with resource assessment to operational maintenance. For example, sensor-laden buoys and subsea drones can gather ongoing oceanographic data (such as wave height, tidal flow, and salinity gradient) that can be fed into predictive models that allow design and operational considerations useful for optimal site selection and energy extraction (e.g., Neumann et al., 2015). Likewise, IoT included in offshore wind or tidal turbines can sense early signs of the mechanical stress, corrosion, or biofouling and may provide condition-based maintenance rather than reactive maintenance for mechanical errors. This predictive method is much more cost-effective and reliable (Figure 8.1).

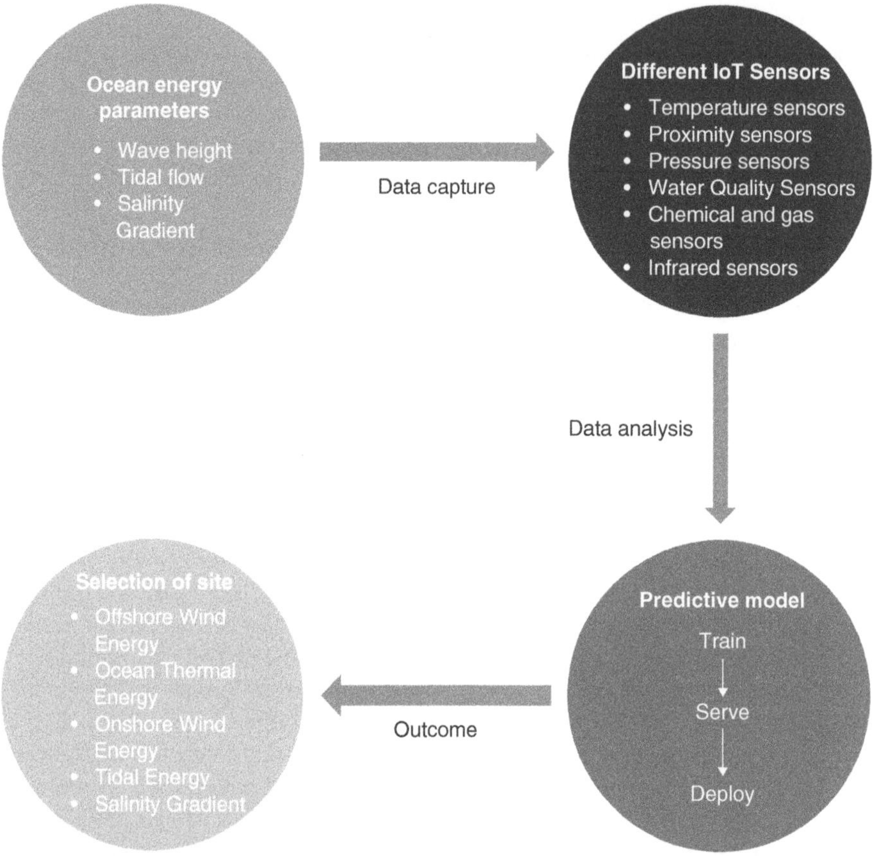

FIGURE 8.1 Process flow chart of the selection of site for the development of effective marine energy infrastructure by the application of IoT.

The IoT also enables to improve the safety and environmental sustainability of maritime operations. Offshore wind farm structures operate in harsh environments with storms, seawater crustaceans and have to withstand tremendous mechanical loads. IoT applications can enable structural health monitoring and avert the risks before catastrophic failures happen (Wang et al., 2021). In addition, environmental monitoring can be supported by electronic observations of their impacts on marine biodiversity, fish migration, or even sediment transport challenging that the renewable energy projects should be consistent with the environmental sustainability. This duality of efficiency and environmental stewardship places IoT as an integral part of responsible marine energy development.

Another disruptive aspect the IoT can bring to marine energy is what it contributes to smart connected offshore grids. IoT connects disparate energy resources such as wind farms, tidal barrages, and floating wave devices into an interconnected offshore energy grid (Hossain et al., 2019). Such networks have the ability to flex in and out of power at varying shifts, and to stockpile excess power through offshore storage, such as underground compressed air or subsea batteries. This vision is of a distributed, smart energy system, and is in harmony with worldwide trends for Industry 4.0 and a digital economy in which data-driven optimization is an inevitable force for competitiveness and sustainability.

IoT promotes human–machine cooperation in the field of marine energy. The integration of IoT with Artificial Intelligence (AI) and Machine Learning (ML) is expected to offer offshore systems an opportunity of becoming autonomous in taking decision. An AI-driven drone can, for example, based on IoT data, inspect offshore platforms, while ML could predict the wave energy conversion efficiency for different weather conditions (Liu et al., 2021). These functions will serve to increase operational resilience as well as to reduce human risk in dangerous marine environments. And with advances in IoT already playing out, full autonomy at offshore energy farms will not seem so far away, transforming the role that marine renewables play in the world's move to cleaner energy.

8.3 IOT ARCHITECTURE FOR OCEAN ENERGY SYSTEMS

In order to further these, it is important for relevant scientific and engineering communities to call for timely technological advances in sensing, monitoring, and communication technologies for the successful deployment of MRE systems. Real-time monitoring has placed sensors and smart meters at the centre of energy generation, operational behaviour, and the environmental status of offshore platforms. In MRE applications like wave and tidal farms, hydrodynamic parameters including current velocity, wave height, and turbulence are monitored using sensors to predict power generation efficiency (Jiang et al., 2022). Smart meters enhance this service by allowing two-way interactions between the energy generators/producers and the consumers, providing granular details at the grid level and facilitating flexible demand management techniques (Chakraborty et al., 2020). When incorporated within marine energy schemes, these tools enable energy system operators to optimize generation schedules, reduce losses, and improve grid integrity (Figure 8.2).

Subterranean communication networks serve to broaden the potential of systems for monitoring and controlling ocean energy. Traditional means of communication such as Radio frequency communication do not work well underwater because of high signal loss. Instead acoustic, optical, and hybrid communication technologies are employed to connect the underwater sensors, autonomous underwater vehicles (AUVs), and buoys (Akyildiz et al., 2021). Long-range connectivity can meanwhile

FIGURE 8.2 Role of IoT service in the improvement of marine renewable energy infrastructure.

be established with acoustics which is however much slower while one can achieve high data rates over spatially short distances using optics. These networks ensure a constant stream of information from offshore energy installations to central control stations that lead to early detection of irregularities and remote intervention. The creation of robust underwater communication networks is essential for scaling marine renewable projects, as they allow for timely decisions, predictive maintenance, and environmental monitoring.

The installation of IoT devices in marine energy systems provides abundant amounts of data that need advanced treatment and storage materials, and also data analysis platforms. Cloud computing has emerged as a valuable solution for managing such data-heavy tasks, where storage and computing potentials are very high at a scalable level. In the maritime/offshore energy sector, cloud platforms can enable the integration of heterogeneous datasets, including turbine vibration and seabed mapping, along with weather forecasts and marine ecosystem audits, into integrated analytics (Goudarzi et al., 2020). With cloud-based analytics, operators can utilize sophisticated algorithms to identify trends, optimize performance, and predict energy production, resulting in improved efficiency and availability.

However, the niche of centralized cloud computing remains challenging in marine environments, in view of the problems of latency, bandwidth, and link drop-outs. These challenges can be addressed by handling data closer to the source, being extensive on the offshore platform or in the local IoT gateways. This decentralized computing approach decreases delays and the reliance on potentially unstable communication channels, and it allows for uninterrupted operation when the connectivity is degraded (Shi et al., 2016). For instance, an edge-enabled marine turbine can be self-aware of possible structural vibrations that may be the prelude to device failure and can then initiate preventative measures without immediate use of far-away cloud servers. This best-of-both-worlds hybrid solution – of cloud's scalability and edge's snappiness – bridges the gap towards a holistic solution that can tackle the complexity of marine data analytics.

The combination of cloud and edge computing also permits more sophisticated applications such as digital twins for MRE systems. These digital twins depend on edge-processed real-time sensor data kept in the cloud to emulate live operations as well as predict system behaviour. Stretching beyond the connection of the physical and the digital worlds, cloud-edge ecosystems enable marine operators to maintain, lower costs, increase safety, and extend the lifetime of assets (Dai et al., 2020). Standalone or coupled functionalities are essential for scaling MRE as a competitive driver of global energy transitions.

The trend towards the decentralization of energy networks is a key element of the future power system, and IoT-based smart grids are the technological bridge. Conventional centralized power systems cannot accommodate the variability and decentralized nature of MRE. Smart grids, which have communication and control systems based on IoT technology, enable not only two-way flows, but also real-time monitoring and dynamic reconfiguration of power distribution (Gharavi & Ghafurian, 2011). By integrating smart meters and intelligent control (IC) units into offshore grids, energy generation can be flexibly coordinated with load flow, so that system security is maintained even with the fluctuation of marine generation.

A major benefit of IoT-based smart grids is the possibility of decentralized energy distribution. Rather than sending all power via a centralized transmission network, decentralized grids enable local communities, industry, and coastal towns to consume renewable electricity produced locally. This model minimizes transmission loss, promotes energy security, and enhances robustness against disturbances (Fang et al., 2012). On the one hand coastal areas can incorporate windfarms offshore (offshore wind), devices to convert the energy of the tides (tidal converters), and devices to harness wave energy into local networks that run semi-independently from the national grid. The IoT makes sure these many assets work together to balance the load and support frequency regulation, and even storage.

Furthermore, the IoT-based smart grids pave the way for a new market will be mechanisms, i.e., all P2P (peer-to-peer) energy trading. Within this system, prosumers (who both produce and consume power) can trade excess marine-based generated power with other prosumers in a P2P fashion utilizing blockchain-based transaction systems (Mengelkamp et al., 2018). Facilitating such decentralized, IoT-driven ecosystems, now coastal communities themselves can actively participate in the energy transition – they can own and manage their own renewable energy resources, and no longer rely on fossil fuels.

Finally, IoT-enabled smart grids provide a common ground for integrating energy storage systems, electric vehicles, and demand-response programs into marine renewable networks. IoT-enabled predictive analytics can attract the surplus energy in storage units (e.g., batteries or compressed air systems) and discharge the stored energy when demanded or when generation is less than instantaneous demand (Lund et al., 2015). This flexibility results in a secure and stable electricity supply, making the intermittence of marine energy a tractable system feature.

8.4 TECHNICAL INNOVATIONS AND APPLICATIONS

Technological developments for efficiency, reliability, and scalability have driven the development of MRE. Marine conditions are challenging and known for their aggressive nature which is why there's a demand for high-quality instruments that'll not only last but perform well. To tackle these challenges, novel technical advancements in energy harvesting systems, structural design, monitoring systems, and their interaction with smart energy infrastructures have been developed by academia and industry. These advances not only widen the prospects of utilization of marine resources but also pave the way for potential applications in broader socio-economic landscapes (such as coastal electrification, desalination, integration to smart grids, etc.).

Floating offshore wind turbines are one of the most notable developments. Unlike conventional fixed-bottom structures, floating platforms make it possible to harness winds in deeper water, where wind is more abundant and reliable. For example, spar-buoy and tension-leg platforms and semi-submersible concepts have been tested and proven in industrial applications like the Hywind Scotland project, which showed that it is already possible to develop economically feasible floating Offshore Feasible Floating Windfarms (Butterfield et al., 2007). These novel platforms not only extend the geographical reach of offshore wind power, but also mitigate visual and environmental effects in coastal areas.

Moreover, a significant development occurs in wave energy converters (WECs). They have shifted from proof of concept prototypes to robust and modular devices. Different concepts of WECs are, for instance, point absorbers, oscillating water columns, and attenuators, all of these being based on different ways of utilizing the kinetic and potential energy of the ocean waves. New updates in adaptive control algorithms enable WECs to adapt its operation modes according to sea states and contribute to the performance improvements and the load reduction (Li & Belmont, 2014). These are important steps towards the realization of commercial wave energy, a sustainable but technically and economically challenging concept for energy generation.

There has also been considerable development in tidal energy systems, particularly in turbine technology. Both horizontal and vertical-axis turbines have been developed to maximize the power extraction from different velocity regimes of tidal flow, as well as ducted and venturi-enhanced approaches. The development of new materials, including biofouling- and fatigue-resistant composites, has prolonged the life of the devices (Polagye et al., 2011). And modular tidal arrays are being developed to get the most power without being too disruptive for the environment. The regularity of tides facilitates their incorporation into the grid system, and new technologies with lower installation and maintenance costs have allowed more widespread exploitation of these resources.

Hybrid systems are not only bringing new generation technologies, but also new fields of marine renewable applications. Hybrid offshore structures that capture various types of energy resources (e.g., wind turbines and wave converters or solar panels), offer higher reliability and more efficient total energy generation power (Borg et al., 2018). The platforms can have built-in energy storage systems to provide stable power when marine resources are variable. Hybrid systems are also representative of the general development trend for integrated, robust, and multi-purpose offshore systems.

A second key development has been the use of sophisticated materials and coatings. Such marine equipment lasts much longer than in days of old, protected by corrosion-resistant alloys, composite blades, and materials that can self-heal insults from seawater. Anti-fouling surface coatings, which draw their inspiration from the living world, assist in ameliorating biofouling, a very significant challenge promoting drag and inefficiency in submerged parts (Salta et al., 2013). And these innovations in materials reduce maintenance costs and overall system efficiency, making marine energy technologies more competitive with other renewables.

Technical developments have also been made with regard to digitalization and monitoring, in particular integration of ML and AI for predictive maintenance and optimization. Sensors and cameras are increasingly mounted onto AUVs or remotely operated vehicles (ROVs) to inspect subsea cables, turbine foundations, and wave energy devices. These digital disruptions lower the risk for humans, increase the accuracy of data, and secure tracking of assets. Predictive analytics can determine when components might fail, how to maximize the energy conversion in the presence of different sea states, and how to guarantee environmental compliance, all of which would allow MRE projects to be economically viable.

The uses of technical innovations are not confined to electricity production. Marine power energy may have a potential role in desalination and freshwater

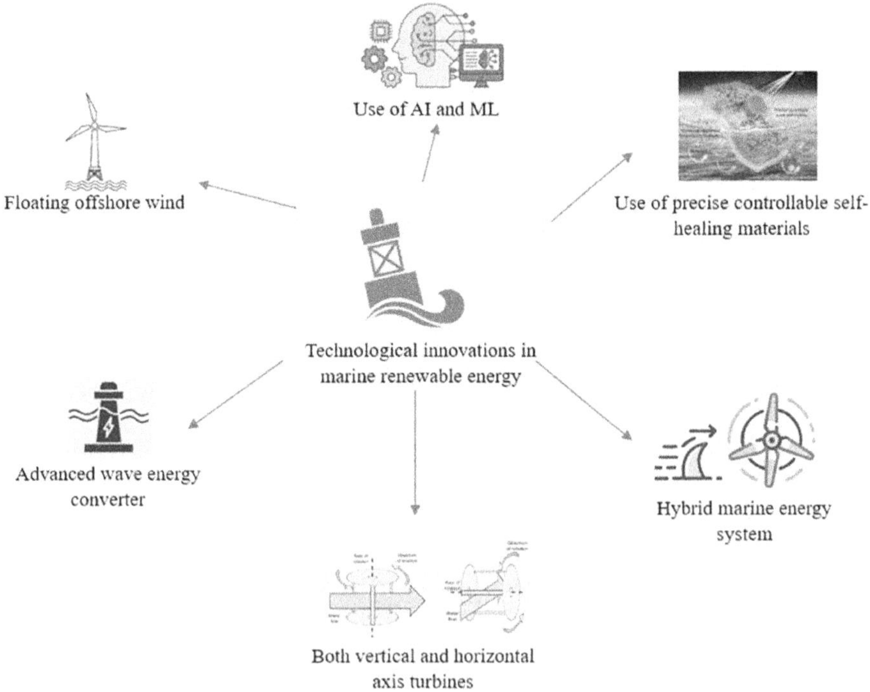

FIGURE 8.3 Technological innovations in marine renewable energy.

production, especially in arid coastal zones. When used in conjunction with wave or tidal energy, reverse osmosis desalination plants offer a sustainable substitute to fossil fuel-driven desalination facilities (Paparella et al., 2019). It should also be added that offshore energy plants are increasingly considered for hydrogen production by electrolysis. Marine source-derived green H2 can play an important role as a clean fuel for transport and industry, and thus contribute to improving energy security and advancing decarbonization pathways (Gaudern et al., 2018).

A further developing application is the integration of ocean renewables with offshore aquaculture and the coastal industry. Hybrid energy systems can provide power to address the needs of fish farms, cold storage facilities, marine research stations, etc., facilitating the development of blue economy on a sustainable basis. By combining renewable energy with offshore economic activities, innovations help move the economy away from dependency on fossil fuels and develop sustainable coastal areas (Buck & Langan, 2017; Figure 8.3).

8.5 LEGAL AND REGULATORY CONSIDERATIONS

The development of MRE is facing specific legal and regulatory challenges, which can be distinguished from those encountered in onshore energy production. Oceans offer a tremendous opportunity for renewable energy from offshore wind, wave, and tidal, but developing these resources is governed by a complex legal regime that seeks to strike a balance among them with national energy policies, environmental protection, navigation, and resource rights. Correct regulatory certainty would drive investment, maintain compliance with environmental legislation, and support the sustainable growth of marine energy. Sound governance is required to address jurisdictional issues (in particular transport), procedures for permitting and licensing, environmental effects, and interaction with national and international legal orders.

One of the most fundamental legal questions in MRE is the decision-making in the sovereign rights of the ocean area. International law of the sea, as regulated by the United Nations Convention on the Law of the Sea (UNCLOS), is the general legal framework for the exploitation of sea resources. Sovereign states have exclusive rights to search and exploit resources in their Exclusive Economic Zone (EEZ), which stretches up to 200 nautical miles offshore (United Nations, 1982). This basis of law permits states to permit the construction of offshore wind or tidal farms in their EEZ, while respecting the international navigation freedom principles. Elsewhere the legal basis for jurisdiction is less obvious, and the result is unclear – who governs MRPs in international waters? Inadequate transnational cooperation mechanisms have been a limiting factor for some projects near maritime borders or in common ecosystems.

The granting of permits and licences at the national level is one of the major regulatory barriers to MRE deployment. A common challenge to the developers was the bureaucracy of a regulatory system that was often split among various agencies in charge of environmental protection, energy planning, and fisheries and marine transport. For instance, in the United States, the Bureau of Ocean Energy Management (BOEM) is responsible for offshore renewable leasing, with other agencies responsible for environmental analysis and grid integration (Firestone et al., 2012). Complex, lengthy, and overlapping processes can discourage investment and delay projects from being developed. Improving licensing regimes, creating one-stop permitting systems, and demonstrating transparent decision-making processes are all broadly seen as essential for the growth of MRE.

Environmental laws are key to the regulation of MRE, as the potential harm to the environment, biodiversity, and fish life makes it necessary to control and regulate renewable energy. Developers must carry out comprehensive studies to measure negative environmental impacts, including noise, disturbance to habitat, and the risk of collision between fauna and the structure under frameworks, such as the European Union's Environmental Impact Assessment (EIA) (Copping et al., 2016). Precautionary evaluation can inform these assessments, as ecotoxicological risks are considered significant enough to warrant risk management measures. These vary from in-linked EIAs covering larger areas making multiple projects at a regional scale in a shared marine area, to the more strategic level assessments such as Strategic Environmental Assessments, in particular at the scale of the region. However, these requirements do need to be considered against the pressing need to speed up renewable energy deployment in line with climate goals.

A related legal dimension is marine spatial planning (MSP), which organizes various forms of ocean space use (shipping, fishing, conservation, and energy generation) using comprehensive frameworks. For example, the EU MSP Directive requires member states to develop national spatial plans that designate sea areas for renewable energy development in ways that help mitigate potential conflict with other maritime uses (Jay et al., 2016). In addition to minimizing disputes within the law, MSP promotes investor confidence through the establishment of priority areas or zones for MRE projects. Similar methods are being developed throughout the world, emphasizing the importance of planning at a spatial level as a legal tool for juggling interests in marine areas.

Spatial and environmental aspects, as well as grid and market regulations, are equally important. Those interested in developing offshore energy need legal instruments for grid connection as well as for transmission rights and pricing structures to ensure fair access to the market. In Europe, offshore grid expansion is steered by the EU's Renewable Energy Directive and Ten-Year Network Development Plan, which promote international interconnection between markets for the integration of offshore wind power (Dedecca et al., 2019). Lack of regulation on offshore grid facilities is still a key hurdle for marine energy technology development in developing countries. Supportive feed-in tariffs, power purchase agreements, and market incentives can provide stable investment climates.

The legal framework also consists of insurance, liabilities, and decommissioning responsibilities. Risks of accidents, environmental pollution, and navigation threats are present for offshore installations. Transparent legal rules for defining the liability of the various actors in cases of spills, collision, or structural failures enable the GHG actors to be held accountable (Esteban et al., 2011). In addition, decommissioning requirements mandate that infrastructure is removed, and it becomes the operator's responsibility to remove the infrastructure and dispose of it safely where its life is exhausted, with the aim that facilities are not left to cause a longer-term environmental or navigation risk. Although such legal obligations add to initial costs, they support responsible and sustainable development of marine renewable projects.

Internationally, with the scale of MRE projects increasing, standardization and trans-boundary cooperation are of growing importance. Cross-border grids at sea, for example, in the North Sea, demand harmonization of technical standards, legal procedures, and regulatory regimes among involved states. (Peecock et al., 2025) International guidance such as that from the International Renewable Energy Agency (IRENA) also offers best practice principles for licensing, environmental assessments, and stakeholder engagement. Standardized rules support legal certainty for developers and support the scale of investment required across localized energy markets.

Governments and regulators should ensure that community and stakeholder participation rights are embedded in laws and regulations. Local communities, indigenous groups, and fishing industries frequently express concerns about the effects of offshore ventures on their livelihoods, and on their cultures. Legal structures for consultation and benefit-sharing, such as required by international conventions including the Aarhus Convention, also serve the interests of transparency, fairness, and social acceptability (O'Hagan & Lewis, 2011). Robust participatory regimes not only reduce disputes, but also can be expected to increase the legitimacy of marine renewable developments.

8.6 POLICY PATHWAYS FOR DECENTRALIZED OCEAN ENERGY

Decentralized ocean-based energy could revolutionize the way renewable power is harvested – casting aside massive megaprojects and moving towards technologies that could give islands and coastal towns much more direct access to the power of the waves and the sun hitting the ocean's surface. Such systems use tidal turbines, WECs, and floating PV to provide the grid with distributed electricity with fewer losses in transmission and higher energy resilience. Effective policy pathways to unlock the potential of decentralized ocean energy need to be combined with supporting finance mechanisms, Adaptive governance, Community involvement, Technological innovation, and Environmental and social sustainability. These policy levers can aid in developing a diversified energy portfolio that benefits local economies while working towards global climate goals.

A key policy direction to decentralized ocean-based power is the facilitative enabling environment for local-scale projects. Conventional regulation regimes can however be built on centralized systems and large-scale infrastructure, potentially excluding smaller community systems. One way to address this may be for government to create specific licence categories for small- to mid-scale ocean energy installations, allowing a simplified and less bureaucratic approvals process but still with stringent safety and environmental considerations (Kirby et al., 2021). Efficient permitting and a standardized approach to environmental review would enable both innovators and community cooperatives to try their hand at local ocean energy solutions without being penalized for the economies of scale that are possible in bigger offshore projects.

Another key path is to embed decentralized energy systems within wider energy planning. States' power systems often focus on grid security and scale renewable power, but many small-scale ocean devices can help realize these goals, enhancing energy security and resilience particularly in isolated coastal areas. For example, mixed energy systems that use power from tidal, wave, and offshore wind power sources with battery systems can be promoted in the context of renewable portfolio standards or microgrid programs (Esteban et al., 2012).

Financial structures are key in making decentralized projects happen, as such projects typically do not have the capital support of utility-scale projects: Renewed policies, Green bonds, concessional loans and targeted subsidies. These policy innovations can reduce financial risk for cooperatives and small-scale developers. In addition, revenues can also be assured if the project is eligible for feed-in tariffs or community energy credits that provide a source of income that can assist in attracting investment (Carley & Konisky, 2020). Public Private Partnership (PPP) also offers a mechanism to replicate demonstration pilots at scale and distribute the financial risk among partners. Crucially, mechanisms for financing should be adapted to the specific challenges of small-scale deployment, including high up-front technology costs and constrained credit markets in developing countries.

A decentralized ocean energy policy should also emphasize community involvement and energy justice. Ocean renewable energy could empower coastal and island regions with a decrease in dependence on imported fossil fuels and a move towards energy autonomy at the local level. When there are policies which require stakeholder engagement, community–owned projects, or cooperative endeavours, local acceptance can be encouraged, and the economic gains can be shared (Jenkins et al., 2016). In small island developing states (SIDS), ocean energy microgrids can provide power to households, schools, and hospitals, contributing to social and economic resilience. The fair and inclusive transition to renewable energy requires communities to have decision-making abilities and financial interests in energy projects.

Another crucial policy route is entrepreneurship and research backing. Most of the decentralized ocean energy technologies are still in the pre-commercial or pilot phase of development and still need continuing research, testing and standardization. Governments can create innovation parks, research groups, and demonstration areas, where prototypes go on the market in order to pass the proof of concept (Cruz & Atcheson, 2016). In an open-access, data-sharing environment, the pace of innovation is pushed quickly forward, and lessons learned can be more rapidly applied to new projects. Local manufacture of ocean energy devices is likewise promoted, achieving continuity and complementarity among technology innovation and regional economic growth, particularly in centre of naval industry.

Policies for environmentally sustainable ocean must be thoughtfully crafted to promote distributed ocean energy without harming the marine environment. For instance, small density installations could mitigate the total environmental risks, compared with megaproject, but still need to be monitored for impacts on, for example, habitats and navigational safety. Adaptive governance systems permitting iterative licensing (ek-licensing of projects followed by stringent monitoring/followed by possible expansion) should be the middle ground (Reeve et al., 2011). Here such adaptive governance allows for the advancement of environmental protection with growing technological maturity.

International cooperation is also an exciting policy avenue to accelerate decentralized ocean energy. Technology transfer and resource mapping can be difficult for small-scale projects. Regional cooperation, in the form of regional pooled testing facilities, joint capacity-building strategies, and standardization of testing, for example, can minimize redundancies and expand capacity beyond borders (Cui and Zhao, 2023). For island states in the Pacific or Indian Ocean, regional arrangements that can maximize resources and capacity could facilitate faster decentralization. Connecting these efforts with international initiatives being pursued by institutions like the IRENA creates a coherent relationship between local action and global decarbonization objectives.

The development of informational technology and smart-grid strategies is also crucial to achieving decentralized OTEC systems. Measures that favour advanced metering, demand-side management, and AI-enabled predictive maintenance enable the distributed systems to operate efficiently in the broader energy networks (Makai and Popoola, 2025). Digital twins and IoT-based monitoring tools can increase the efficient operation of ocean energy-run microgrids, providing stability and reliability even at remote locations. Policymakers can facilitate digital infrastructure investment through complementary national energy transition strategies, achieving synergies in technology and policy innovation (Figure 8.4).

8.7 SOCIO-ECONOMIC IMPLICATIONS

Decentralized ocean energy has significant social and economic ramifications, especially in coastal and island areas where access to power and the job market is frequently limited. Such decentralized systems, like tidal turbines, wave converters,

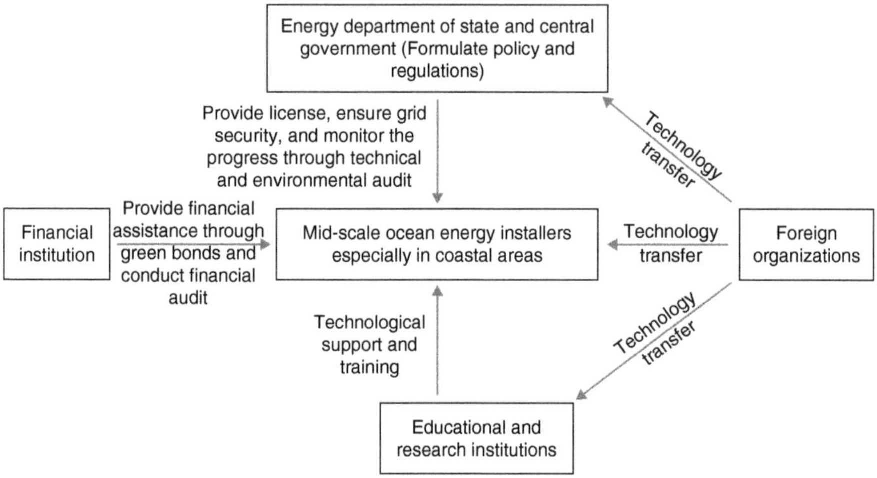

FIGURE 8.4 Framework of the implementation of effective decentralized ocean energy strategy.

and floating solar, are aimed at offering localized energy solutions rather than huge centralized offshore megaprojects. These smaller initiatives can contribute to tackle directly energy poverty, increase resilience, and drive regional development in line with more comprehensive frameworks for sustainable development. The social and economic implications extend beyond the delivery of clean power to impact the incomes, jobs, and overall welfare of communities.

One of the main socio-economic advantages of decentralized ocean energy is the opportunity it offers for improving access to, and security of, energy in far-flung and less well served areas. Many SIDS, for instance, are particularly dependent on costly, imported fossil fuel for their energy consumption, making them vulnerable to fluctuations in global oil prices and supply volatility. Decentralized systems could cut reliance on imports, decrease costs of energy, and lead to more autarky, through using available resources found in the sea locally (Sheng, 2019). Better energy security does more than raise living standards; it puts less fiscal strain on governments that have been generous with fuel subsidies.

Decentralized ocean energy has great potential for local economic development as well. Implementing small and medium-scale projects generates opportunities for local industries in production, assembly, installation, and maintenance. Maritime services, engineering and port operations contribute supply-chain opportunities for coastal communities. Furthermore, through the growth of community-scale energy projects, renewable energy skillsets are increasingly developed, creating a basis for long-term growth of human capital (Magagna & Uihlein, 2015). This economic diversification is especially important for coastal economies that depend otherwise in their fisheries or tourism – sectors that will be at high risk from the effects of climate change.

Another important socio-economic consequence is employment generation. Decentralized energy projects tend to be relatively labour-intensive during construction and installation and also provide relatively long-term employment in operation and maintenance. Tidal and wave energy deployment, for instance, requires marine engineers, technicians, and vessel operators, and can help create jobs in areas where conditions of unemployment are frequently higher (Allan et al., 2015). The ripple effect extends into sectors such as logistics, higher education, and technology transfer, stretching the economic tapestry of coastal communities.

Decentralized ocean energy has a role to play in social equality and inclusion. Some examples are found in the community-based 'ownership or cooperative model' that allows local communities to share in the revenues and, in doing so, reduces inequality and increases social capital. Cheap, reliable electricity also boosts socio-economic opportunities through its ability to power small businesses, run schools, health centres, and libraries, and increase access to digital services (Agarwala, 2025). This local empowerment is also in line with the concept of energy justice, thereby preventing marginalized communities from being left behind in the worldwide switch to renewable energy.

Socio-economic benefits relate as well to climate resilience and adaptation. Maritime and insular areas are especially threatened by sea level rise and extreme climatic events. Library of Congress Decentralized power systems, especially if in the form of a microgrid, can keep the lights on when the centralized grid goes down during a disaster. Access to electricity enables the provision of critical services such as hospitals, ambulance facilities, and desalination plants, vital in post-disaster recovery (Kerr et al., 2020). This facet of resilience has social and economic elements and can reduce the costs for long-term recovery, and it has the effect of also saving human lives.

8.8 CHALLENGES

There are various obstacles to decentralized ocean energy and its widespread uptake and commercialization. These problems, which are technical, economic, environmental, and social in nature, reflect the problems associated with the implementation of renewable-based systems in a marine environment. Overcoming these barriers is a condition for decentralized ocean power to be able to move from being experimental endeavours to reliable forms of sustainable power which coastal and island communities depend on. There are also socio-economic challenges that need to be addressed. The high capital costs of ocean energy technologies may make them out of reach for poorer regions without targeted financing and subsidies. Conflicts may also arise over marine space, especially where fishing communities are dependent on coastal waters (Johnson et al., 2012) Meeting these challenges demands a holistic approach, reconciling technology deployment with stakeholder engagement, fair financing, and marine spatial planning.

Technologically, it is still a bit immature and not reliable. A number of ocean technologies (e.g., small-scale tidal turbines and WECs) are still at the pilot or pre-commercial level. Intense marine environments such as those defined by corrosion, biofouling, and extreme weather can disrupt operations due to a large number of failures and frequent maintenance requirements (Borthwick, 2016). These problems not only lead to higher operating fees, but also ruin investor trust in decentralized projects. Unlike offshore wind, which is relatively mature and has scaled up, our wave and tidal technologies are still in the early stages of research, development, testing, and standardization to be deployable at community scales.

Another key obstacle is financial viability. The capital costs for developing and placing decentralized ocean energy systems are typically greater than those for land-based renewables like solar or wind. There are also costs for the subsea cabling, anchoring, and specialist marine vessels to install the device (Coles et al., 2021). These up-front costs are often not financially feasible for SIDS or rural coastal communities unless subsidies, concessional finance, or international support are provided. Moreover, there are no established markets for ocean energy components, no opportunity to benefit from economies of scale, and costs are therefore high.

There are also grid integration and infrastructure issues that work against locally distributed ocean energy. Many coastal and island communities also have inadequate grid infrastructure that is unable to handle the variability in marine energy. Even with proper storage technologies and microgrid implementation, energy supply could still be vulnerable. Moreover, a poor and inadequate transmission network may limit the spread of ocean energy generation and decrease realization of the potential for decentralized systems (Arcos-Avilés et al., 2018).

The environment and ecology also pose another critical challenge. Though decentralized projects are smaller than large offshore farms, they too pose risks like disturbance to marine environments, noise pollution, and disruption to navigation and fishing. The turbine blade, electromagnetic field surrounding the subsea cable, and the actual presence of devices could all alter the behaviour of marine mammals, seabirds, and fish communities (Copping et al., 2016). These long-term ecological uncertainties result in the added challenges of regulation and project approval delays. We need adaptive environmental monitoring frameworks, but this also adds cost.

From a governance standpoint, regulations are complicated, and no policy exists to support the coordination of decentralized ocean energy. Most national regulatory contexts have been set up to deal with big offshore energy projects, and community-based projects fall into the cracks of policymaking. In fact, lengthy permit procedures, competing jurisdictional claims to marine areas, and the lack of clear criteria regarding decentralized projects can drive developers away (O'Hagan et al., 2017). Small developers or co-ops are hit hardest by this regulatory uncertainty and are simply unable to financially or legally navigate complicated approval processes.

The second significant issue is that of social acceptance and community involvement. The backlash from these has led many localized projects to focus on empowering local people, while the general population tends to ignore these plants and animals or create oxidation resistance if they perceive a potential threat to livelihoods (fishing, tourism). Perceptions of visual intrusion, risk, and possible extant impacts on biodiversity may cause resistance to projects near coastal communities (Devine-Wright, 2011) 11. Transparent consultation, fair benefit-sharing mechanisms, and participatory decision making are essential for strong trust and local support.

The issue of knowledge gaps and capacity building has been addressed in this study. Most coastal and island communities do not have the knowledge and capability to develop and implement ocean energy devices. Decentralized initiatives may become dependent on external experts if training programs and knowledge transfer

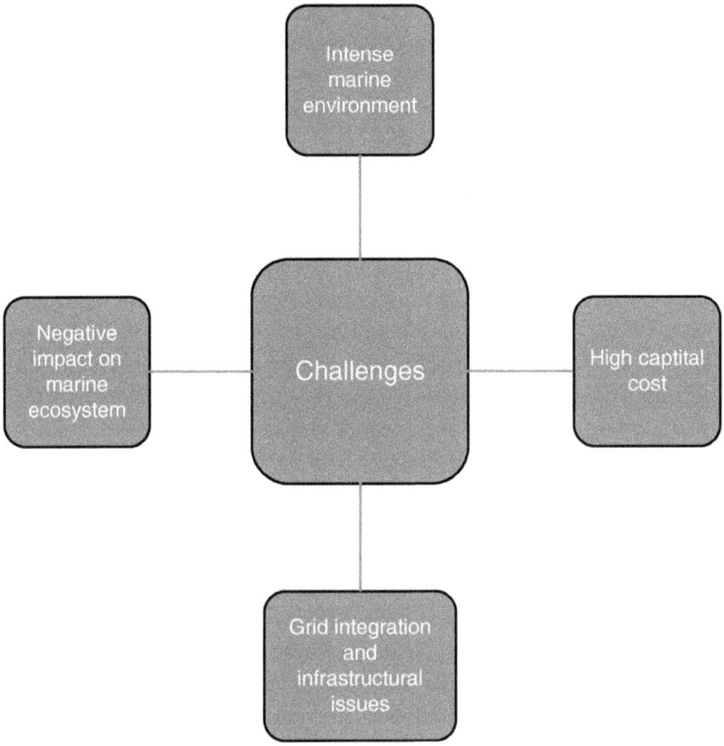

FIGURE 8.5 Challenges of the installation of marine renewable energy system.

are not pursued, jeopardizing the sustainability and the local ownership (and economic returns) of the projects (Soukissian et al., 2017). Consequently, the development of institutional and human capacity can be just as crucial as the technological development (Figure 8.5).

8.9 FUTURE PERSPECTIVE

The promise of ocean energy in the future of a decentralized system is huge as the world evolves towards sustainable, reliable, and community-based energy solutions. Amid current global dilemmas concerning clean energy supply and climate change abatement, distributed ocean energy presents an alternative approach that connects technological innovation with social agency and ecological care. In possible futures, advances in engineering, digitalization, policy frameworks, and international cooperation are expected to define the development of the sector, collectively reconfiguring how marine resources are utilized.

One key focus going forward will be continued progression in the technology. Marine technology is undergoing rapid transformation from experimental prototypes to industrial and commercial applications. The novel floating tidal turbines, modular wave converters, and integrated hybrid platforms have become more efficient, reliable, and versatile for applications around the world. As these materials develop, both near-resistance to corrosive and biofouling needs of maintenance will be decreased, and IC systems will be used to enhance the power capture in a timely fashion. These advances will reinforce the reliability and cost-effectiveness of decentralized systems, ensuring that they achieve a higher rate of acceptance in coastal and island areas where a district grid is less feasible.

The digitalization and smart energy management at decentralized ocean systems are just as crucial. Digital twins, AI, and predictive analytics will be critical to improve operational efficiency and minimize downtime. Digital technology will provide for real-time monitoring of underwater equipment thereby preserving best–in-class performance and minimizing environmental risk. Combining decentralized ocean energy with microgrids and blockchain-based trading platforms will also enable the autonomous management of energy within the communities, thus generating opportunities for economic empowerment and resilience in the light of worldwide uncertainty with regard to energy matters.

The social dimension of decentralized ocean energy in the future is closely related to community involvement and engagement. Smaller-scale, community-led projects will create space not only for the expansion of energy democracy – in which communities are not just consumers but also co-owners of local energy systems – but also the potential to reduce the power of incumbent energy and water companies. This model will offer more broad-based benefits in terms of job opportunities, skills training, and local income. As energy justice moves to the fore of the global transition, decentralized ocean energy has the potential to keep vulnerable and marginalized coastal communities as active participants in creating sustainable futures.

Another view focuses on how distributed ocean energy is helping us adapt to our changing climate. Coastal and insular areas are considered some of the most vulnerable to increases in sea level, to extreme weather patterns, and to energy supply-chain disruptions. Decentralized ocean energy systems in the form of resilient microgrids can provide emergency electricity to hospitals, desalination plants, and emergency services in times of disasters. In this way, decentralized energy is not just one tool for decarbonization: it is a lifeline for adaptation and survival amid a changing climate.

On an international scale, there will be increased focus on policy alignment and international cooperation in the future as well. While nations move together towards common climate goals, decentralized ocean energy is likely to be incorporated within collaborative frameworks that incentivise technology transfer, shared testing facilities, and cross-border financing. SIDS, in particular, stand to gain from regional collective action by sharing resources and expertise. International organizations and climate funds International organizations and climate funds will be doing more de-risking of investment, which will make decentralized systems more attractive for both private developers and local cooperatives.

Economically, the future outlook indicates that costs will tend to decrease gradually as technologies mature and economies of scale are achieved. The emerging offshore renewables market, and especially for offshore wind, will introduce economies of scale in production, logistics, and maritime services, which could be efficiently exploited by scattered installations. At the same time, novel financing structures like green bonds, impact investing and community cooperatives will unlock additional sources of capital to make decentralized ocean energy part of not only wealthy communities, but indeed all of the emerging economies as well.

Decentralized ocean energy, in the future, will support multi-use marine infrastructures. If devices can be co-located with aquaculture, desalination, research, and possibly a rich tourist attraction, not only will they make optimal use of ocean space, but they also offer a range of economic opportunities. These multiple-use systems will represent a sea change in ocean governance towards the creation of an integrated, environmentally responsible blue economy, in which energy production is integrated with food supply, water management, and livelihood improvement.

8.10 CONCLUSION

Decentralized ocean energy is a game-changing route to a sustainable, resilient, and equitable energy future. Derived from a variety of marine renewable resources, including tide, wave, and offshore wind, these systems represent the heights of technical innovation while at the same time supporting global climate and development objectives. The use of digital technologies such as IoT, smart grids, and advanced analysis guarantees the efficiency, robustness, and broad-spectrum elasticity in harsh maritime conditions. Simultaneously, rulebooks and environments need to ensure environmental security, safety, and access, while policy routes need to emphasize local governance, international collaboration, and financial innovation to open investment. Outside of the technical and policy realms, decentralized ocean energy has deep socio-economic implications. It is a job creator, gives power to the people, and is a great equalizer, especially in the hundreds of remote coastal and island communities. But stumbling blocks such as the high price of investment, teething technological problems, and environmental concerns highlight the vital importance of ongoing innovation and flexible rule-making. Decentralized systems will be an integral way to provide clean energy and build climate resilience as part of comprehensive blue economy approaches. Looking ahead, decentralized systems won't just provide clean energy; they will help shore up climate resilience and help inform integrated blue economy strategies. In short, decentralized ocean energy can play a unique role in connecting local empowerment with global sustainability, providing a crucial nexus of future energy transitions.

REFERENCES

Akyildiz, I. F., Pompili, D., & Melodia, T. (2021). Underwater acoustic sensor networks: Research challenges. *Ad Hoc Networks*, 34, 299–315. https://doi.org/10.1016/j.adhoc.2015.01.001

Allan, G., Gilmartin, M., McGregor, P., & Swales, K. (2015). The regional economic impacts of renewable energy–A Scottish analysis. *Energy Policy*, 78, 544–555. https://doi.org/10.1016/j.enpol.2014.12.023

Arcos-Avilés, D., Sigrist, L., & Morari, M. (2018). Challenges and opportunities in microgrid integration of ocean renewable energy. *Renewable and Sustainable Energy Reviews*, 96, 379–391. https://doi.org/10.1016/j.rser.2018.07.044

Atzori, L., Iera, A., & Morabito, G. (2010). The internet of things: A survey. *Computer Networks*, 54(15), 2787–2805. https://doi.org/10.1016/j.comnet.2010.05.010

Agarwala, N. (2025). Marine Renewable Energy and Its Role in Sustainable Blue Economy. In: Leal Filho, W., Salvia, A.L., Eustachio, J.P.P., Dinis, M.A.P. (eds) *Handbook of Sustainable Blue Economy*. Springer, Cham. https://doi.org/10.1007/978-3-031-32671-4_34-1

Borg, M., Collu, M., & Brennan, F. (2018). Offshore renewable hybrid energy systems: Current status and future perspectives. *Ocean Engineering*, 147, 260–275. https://doi.org/10.1016/j.oceaneng.2017.10.048

Borthwick, A. G. L. (2016). Marine renewable energy seascape. *Engineering*, 2(1), 69–78. https://doi.org/10.1016/J.ENG.2016.01.011

Buck, B. H., & Langan, R. (2017). *Aquaculture Perspective of Multi-use Sites in the Open Ocean: The Untapped Potential for Marine Resources in the Anthropocene*. Springer.

Butterfield, S., Musial, W., Jonkman, J., & Sclavounos, P. (2007). *Engineering Challenges for Floating Offshore Wind Turbines*. National Renewable Energy Laboratory (NREL).

Cui, Y. & Zhao, H. (2023). Marine renewable energy project: The environmental implication and sustainable technology. *Ocean & Coastal Management*, 232, 106415, https://doi.org/10.1016/j.ocecoaman.2022.106415.

Esteban, M., & Leary, D. (2012). Current developments and future prospects of offshore wind and ocean energy. *Applied Energy*, 90(1), 128–136. https://doi.org/10.1016/j.apenergy.2011.06.011

Falcão, A. F. O. (2010). Wave energy utilization: A review of the technologies. *Renewable and Sustainable Energy Reviews*, 14(3), 899–918. https://doi.org/10.1016/j.rser.2009.11.003

Carley, S., & Konisky, D. M. (2020). The justice and equity implications of the clean energy transition. *Nature Energy*, 5(8), 569–577. https://doi.org/10.1038/s41560-020-0641-6

Chakraborty, S., Bhattacharya, A., & Kumar, R. (2020). Smart metering infrastructure for smart grid applications: A review. *Renewable and Sustainable Energy Reviews*, 52, 1362–1372. https://doi.org/10.1016/j.rser.2015.07.068

Coles, D. S., Blunden, L. S., & Spence, R. (2021). The economics of wave and tidal energy: Challenges and opportunities. *Renewable Energy*, 169, 967–981. https://doi.org/10.1016/j.renene.2021.01.046

Copping, A., Hanna, L., Van Cleve, B., Blake, K., Anderson, R., & Sather, N. (2016). Environmental risk evaluation system—an approach to ranking risk of ocean energy development on animals and ecosystems. *Human and Ecological Risk Assessment: An International Journal*, 22(3), 593–610. https://doi.org/10.1007/s12237-014-9816-3

Cruz, J., & Atcheson, M. (2016). Deployment of wave and tidal energy devices: Observations and lessons learned. *Renewable and Sustainable Energy Reviews*, 58, 1201–1211. https://doi.org/10.1016/j.rser.2015.12.143

Dai, H. N., Wang, H., Xu, G., Wan, J., & Imran, M. (2020). Big data analytics for manufacturing internet of things: Opportunities, challenges and enabling technologies. *Enterprise Information Systems*, 14(4), 1–25. https://doi.org/10.1080/17517575.2019.1633689

Devine-Wright, P. (2011). Public engagement with large-scale renewable energy technologies: Breaking the cycle of NIMBYism. *Wiley Interdisciplinary Reviews: Climate Change*, 2(1), 19–26. https://doi.org/10.1002/wcc.89

Dedecca, J. G.; Hakvoort, R. A.; Herder, P. M. (2019). The integrated offshore grid in Europe: Exploring challenges for regional energy governance. *Energy Research & Social Science*, 52, 55–67; https://doi.org/10.1016/j.erss.2019.02.003

Esteban, M. D., López-Gutiérrez, J. S., Negro, V., & Malumbres, M. (2011). Methodology for the assessment of the risk levels of maritime structures for extreme events. *Reliability Engineering & System Safety*, 96(8), 1007–1019. https://doi.org/10.1016/j.ress.2011.03.001

Esteban, M., Leary, D., & Asaeda, T. (2012). Current developments and future prospects of offshore wind and ocean energy. *Applied Energy*, 90(1), 128–136. https://doi.org/10.1016/j.apenergy.2011.06.011

Fang, X., Misra, S., Xue, G., & Yang, D. (2012). Smart grid — The new and improved power grid: A survey. *IEEE Communications Surveys & Tutorials*, 14(4), 944–980. https://doi.org/10.1109/SURV.2011.101911.00087

Firestone, J., Kempton, W., Lilley, M. B., & Samoteskul, K. (2012). Public acceptance of offshore wind power across regions and through time. *Journal of Environmental Planning and Management*, 55(10), 1369–1386. https://doi.org/10.1080/09640568.2012.682782

Gaudern, N., Harrison, R., & Walker, J. (2018). Offshore renewable energy and hydrogen production: Opportunities and challenges. *International Journal of Hydrogen Energy*, 43(33), 15533–15549. https://doi.org/10.1016/j.ijhydene.2018.06.084

Gharavi, H., & Ghafurian, R. (2011). Smart grid: The electric energy system of the future. *Proceedings of the IEEE*, 99(6), 917–921. https://doi.org/10.1109/JPROC.2011.2124210

Goudarzi, M., Ilager, S., & Buyya, R. (2022). Cloud Computing and Internet of Things: Recent Trends and Directions. In: Buyya, R., Garg, L., Fortino, G., Misra, S. (eds), *New Frontiers in Cloud Computing and Internet of Things. Internet of Things*. Springer, Cham. https://doi.org/10.1007/978-3-031-05528-7_1

Hahn, B., Durstewitz, M., & Rohrig, K. (2020). Reliability of wind turbines: Experiences of 15 years with 1500 WTs. *Wind Energy*, 13(1), 9–26. https://doi.org/10.1002/we.408

Hossain, M. S., Madlool, N. A., Rahim, N. A., Selvaraj, J., Pandey, A. K., & Khan, A. F. (2019). Role of smart grid in renewable energy: An overview. *Renewable and Sustainable Energy Reviews*, 60, 1168–1184. https://doi.org/10.1016/j.rser.2015.09.078

International Renewable Energy Agency (IRENA). (2021). *World Energy Transitions Outlook 2021*. IRENA.

Jay, S., Alves, F. L., O'Mahony, C., Gomez, M., Rooney, A., Almodovar, M., … & Wang, Z. (2016). Transboundary dimensions of marine spatial planning: Fostering inter-jurisdictional relations and governance. *Marine Policy*, 65, 85–96. https://doi.org/10.1016/j.marpol.2015.12.025

Jenkins, K., McCauley, D., Heffron, R., Stephan, H., & Rehner, R. (2016). Energy justice: A conceptual review. *Energy Research & Social Science*, 11, 174–182. https://doi.org/10.1016/j.erss.2015.10.004

Jiang, Y., Li, D., & Yang, S. (2022). A review of ocean energy resource assessment and prediction technology. *Renewable Energy*, 190, 947–963. https://doi.org/10.1016/j.renene.2022.03.064

Johnson, K., Kerr, S., & Side, J. (2012). Marine renewables and coastal communities—Experiences from the UK. *Ocean & Coastal Management*, 66, 9–14. https://doi.org/10.1016/j.ocecoaman.2012.04.002

Kerr, S., Watts, L., Colton, J., Conway, F., Hull, A., Johnson, K., Jude, S., Kannen, A., MacDougall, S., McLachlan, C., & Potts, T. (2020). Establishing an agenda for social studies research in marine renewable energy. *Energy Policy*, 119, 697–703. https://doi.org/10.1016/j.enpol.2018.04.017

Khan, J., Bhuyan, G., Iqbal, M., & Quaicoe, J. (2009). Hydrokinetic energy conversion systems and assessment of horizontal and vertical axis turbines for river and tidal applications: A technology status review. *Applied Energy*, 86(10), 1823–1835. https://doi.org/10.1016/j.apenergy.2009.02.017

Kirby, M. E., Reilly, K., & Chapman, R. (2021). Regulating marine renewable energy: Opportunities and challenges. *Energy Policy*, 149, 112022. https://doi.org/10.1016/j.enpol.2020.112022

Makai, L and Popoola, O. (2025). A review of micro-hybrid energy systems for rural electrification, challenges and probable interventions. *Renewable Energy Focus* 54, 100687, https://doi.org/10.1016/j.ref.2025.100687

Lewis, A., Estefen, S., Huckerby, J., Musial, W., Pontes, M. T., & Torres-Martinez, J. (2011). Ocean energy. In O. Edenhofer et al. (Eds.), *IPCC Special Report on Renewable Energy Sources and Climate Change Mitigation* (pp. 497–534). Cambridge University Press.

Li, L., & Belmont, M. R. (2014). Model predictive control of a full-scale wave energy converter. *Renewable Energy*, 72, 160–171. https://doi.org/10.1016/j.renene.2014.07.010

Liu, Y., Yu, Y., Wang, H., & Zhou, Q. (2021). Machine learning for marine renewable energy: A review. *Renewable and Sustainable Energy Reviews*, 138, 110649. https://doi.org/10.1016/j.rser.2020.110649

Lund, H., Østergaard, P. A., Connolly, D., & Mathiesen, B. V. (2015). Smart energy and smart energy systems. *Energy*, 68, 1–11. https://doi.org/10.1016/j.energy.2014.02.089

Magagna, D., & Uihlein, A. (2015). Ocean energy development in Europe: Current status and future perspectives. *International Journal of Marine Energy*, 11, 84–104. https://doi.org/10.1016/j.ijome.2015.05.001

Mengelkamp, E., Gärttner, J., Rock, K., Kessler, S., Orsini, L., & Weinhardt, C. (2018). Designing microgrid energy markets: A case study: The Brooklyn Microgrid. *Applied Energy*, 210, 870–880. https://doi.org/10.1016/j.apenergy.2017.06.054

Neumann, F., Callies, U., & Groll, N. (2015). Marine renewable energy: Resource characterization and physical effects. *Journal of Renewable and Sustainable Energy*, 7(1), 013115. https://doi.org/10.1063/1.4905549

O'Hagan, A. M., & Lewis, A. W. (2011). The existing law and policy framework for ocean energy development in Ireland. *Marine Policy*, 35(6), 772–783. https://doi.org/10.1016/j.marpol.2011.01.006

O'Hagan, A. M., de Andrade, F., & O'Callaghan, J. (2017). Regulatory frameworks for marine renewable energy: A global review. *Ocean & Coastal Management*, 135, 1–10. https://doi.org/10.1016/j.ocecoaman.2016.11.005

Paparella, F., Pisanò, F., & Calise, F. (2019). A review on renewable-powered desalination: Technologies, opportunities, and challenges. *Desalination*, 461, 48–67. https://doi.org/10.1016/j.desal.2019.03.006

Peecock, A., Huang, J., Martínez-Felipe, A., & McKenna, R. C. (2025) Reviewing sector coupling in offshore energy system integration modelling: The North Sea context. *Renewable and Sustainable Energy Reviews*, 210, 115220; https://doi.org/10.1016/j.rser.2024.115220

Polagye, B., Van Cleve, B., Copping, A., & Kirkendall, K. (2011). *Environmental Effects of Tidal Energy Development*. U.S. Department of Commerce, NOAA Technical Memorandum.

Reeve, D. E., Chadwick, A. J., & Fleming, C. A. (2011). Coastal engineering and regulation: Toward adaptive management. *Coastal Engineering*, 58(6), 447–459. https://doi.org/10.1016/j.coastaleng.2010.12.006

Salta, M., Wharton, J. A., Blache, Y., Stokes, K. R., & Briand, J. F. (2013). Marine biofilms on artificial surfaces: Structure and dynamics. *Environmental Microbiology*, 15(11), 2879–2893. https://doi.org/10.1111/1462-2920.12186

Shi, W., Cao, J., Zhang, Q., Li, Y., & Xu, L. (2016). Edge computing: Vision and challenges. *IEEE Internet of Things Journal*, 3(5), 637–646. https://doi.org/10.1109/JIOT.2016.2579198

Sheng, W. (2019). Wave energy conversion and its socio-economic benefits for island communities. *Renewable and Sustainable Energy Reviews*, 104, 379–389. https://doi.org/10.1016/j.rser.2019.01.027

Soukissian, T. H., Karathanasi, F. E., & Ponce de León, S. (2017). Marine renewable energy in the Mediterranean Sea: Status and perspectives. *Renewable Energy*, 111, 830–844. https://doi.org/10.1016/j.renene.2017.04.025

Tao, F., Qi, Q., Liu, A., & Kusiak, A. (2019). Data-driven smart manufacturing. *Journal of Manufacturing Systems*, 48, 157–169. https://doi.org/10.1016/j.jmsy.2018.01.006

United Nations. (1982). *United Nations Convention on the Law of the Sea (UNCLOS)*. United Nations.

Wang, H., Zhang, J., & Li, Y. (2021). Structural health monitoring in offshore renewable energy: A review. *Renewable and Sustainable Energy Reviews*, 135, 110238. https://doi.org/10.1016/j.rser.2020.110238

9 Supervised Learning-Driven Insights into Charging Behavior of Electric Vehicle Batteries under Changing Thermal Environments

Yogesh Kumar Rathore, Sumegh Tharewal, Kamal Upreti, B. Sivaneasan, and Pravin Kshirsagar

9.1 INTRODUCTION

The rapid advancement of electric vehicle (EV) technology has accelerated global development toward smart and sustainable mobility. Still, intrinsic barriers such as peak thermal management, state of charge (SOC) estimation, and maximum charging strategies are essential to foster EV performance, safety, and longevity. Machine learning (ML) has been a game-changer for overcoming these challenges using data-driven optimization and predictive knowledge. Park and Kim (2019) utilized supervised learning for peak thermal management with enhanced energy efficiency and thermal control in EV systems. Building on this, Shahriar et al. (2020) presented a thorough overview of ML strategies for EV charging behavior, including user patterns modeling and the interaction with the grid. Estimation of SOC and battery management has been further advanced using hybrid and deep learning (DL) approaches in recent research. Mousaei et al. (2024) highlighted the importance of ML for effective SOC management, while Mazhar et al. (2023) compared different ML algorithms to optimize EV charging in smart grid configurations. Naresh (2025) presented heterogeneous ensemble learning for thermal control, showcasing enhanced battery safety and efficiency. In a similar vein, Shahriar et al. (2022) presented a hybrid recurrent learning model with explainable AI for accurate SOC estimation.

Applications of ML in larger energy systems have also been researched. Ur Rehman et al. (2020) optimized photovoltaic-battery-EV systems under real climatic conditions using data analytics, whereas Nieduzak et al. (2026) formulated predictive algorithms for temperature-sensing battery enclosures to improve

DOI: 10.1201/9781003671039-9

operating safety. Dominguez-Jimenez et al. (2020) studied seasonal trends in EV charging by using data science methods, and Nieduzak et al. (2024) used analogous methods for self-sensing EV battery structures. Neural networks and DL have also been employed for battery performance prediction. El-Sayed et al. (2024) and Al-Haddad et al. (2024) employed ML and neural network models to estimate SOC and predict heat flux, respectively. Arandhakar et al. (2022) proposed intelligent bidirectional charging with recurrent neural networks (RNNs) in view of electromagnetic interference and temperature. Kumar et al. (2023) discussed recent trends in battery modeling and management systems, pointing out the synergistic role of ML in battery health monitoring, charge/discharge characteristics, and thermal management.

Integrating physics-based models with data-driven methods further strengthens the predictive accuracy of EV systems. Frendo et al. (2020) established intelligent charging models for mixed fleets, whereas Goswami (2024) introduced an AI-based multiphysics model to predict Li-ion battery thermal failure. Kleiner et al. (2021), and Çolak (2023) showed sophisticated monitoring of thermal conditions and ML-based power electronics forecasting for different driving conditions. Achariyaviriya et al. (2023) forecasted EV energy consumption based on sensor information, and Zahedi et al. (2022) simulated renewable energy integration in cold-climate EV systems. Improved DL-based fault detection and predictive maintenance capabilities are recent advancements. Kosuru and Kavasseri (2023) developed a DL model for battery management system (BMS) sensor fault diagnosis, while Khan et al. (2024) presented an evolutionary intelligence-supported DL solution for precise SOC estimation. Thorough reviews by Renold and Kathayat (2024) and sophisticated sensorless monitoring methods by Zheng et al. (2023) demonstrate the incorporation of digital twins, physics-based analytics, and ML to enhance performance and reliability. Taken as a whole, these studies show that ML is revolutionizing EV technology—ranging from energy management and charging habits to predictive diagnostics and safety validation—allowing for sustainable, intelligent, and self-optimizing electric mobility environments.

Figure 9.1 presents a supervised learning-based methodology for the modeling of EV battery charging behavior. The battery system is shown connected to a charge unit, with sensors measuring voltage, current, temperature, and SOC. Sensor data is input to a supervised learning algorithm, which estimates optimal charging profiles for various thermal states. Feedback loops from model output back to the charging control system are represented by arrows, pointing out the effects of temperature changes on charging efficiency, battery aging, and safety.

The subsequent sections will expand on the current discussion by discussing the survey methodology, including how ML methods can be leveraged to analyze, compare, and evaluate performance with regard to relevant datasets on EVs. Next, the section on results and discussion will provide focal points on the results related to ML innovations in battery management and predictions in maintenance/intelligent operation of EVs to demonstrate key results by comparing types of learning models/other datasets. Finally, the conclusion will provide a synthesis of the technological trends, research gaps, and brief recommendations for future directions to improve the reliability/safety/sustainability of next-generation electric mobility systems.

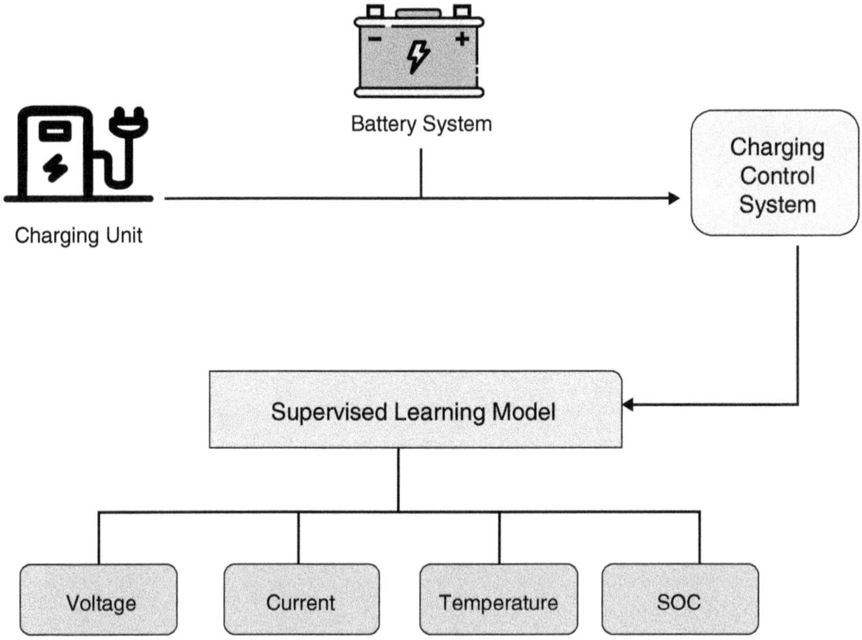

FIGURE 9.1 Schematic of EV battery charging under changing temperature conditions.

9.2 LITERATURE SURVEY

ML techniques have increasingly been applied in EVs' different subsystems with demonstrated benefits in thermal management, charging optimization, SOC estimation, fault detection, and battery health prediction. Supervised-learning-based thermal management techniques offered quantifiable gains in temperature regulation and energy efficiency in battery packs (Park & Kim, 2019). Shahriar et al. (2020) reviewed charging behavior modeling and pointed to ML's ability to learn user behavior and enable grid-interactive policies that enhanced load balancing and reduced peak demand. Mousaei et al. (2024) combined traditional battery modeling with data-driven techniques, which maximized estimation accuracy and responsiveness to varying operating conditions. Mazhar et al. (2023) conducted comparative ML research for smart-grid-integrated charging systems and demonstrated the feasibility of classical and DL models for fleet-level coordination and demand-side management. Shahriar et al. (2022) developed recurrent and hybrid explainable models that enhanced temporal SOC estimation with greater interpretability for BMS stakeholders. El-Sayed et al. (2024) proposed data-driven SOC and state of health (SOH) estimation methods that used convolutional, recurrent, and ensemble models to improve predictive performance while focusing on dataset quality and generalizability. Al-Haddad et al. (2024) utilized neural-network-based finite element or surrogate models to estimate thermal flux and temperature distribution, supporting real-time thermal monitoring and battery design. Arandhakar et al. (2022) introduced recurrent neural network-based bidirectional charging regulators that showed potential for vehicle-to-grid (V2G) coordination, considering electromagnetic interference and thermal effects. Kumar et al. (2023) compiled large-scale reviews incorporating battery

TABLE 9.1
Summary of literature survey

References	Methodology Used	Key Findings	Limitations
Park & Kim (2019)	Supervised learning models for thermal management (regression/classification-based controllers)	Improved energy efficiency and tighter temperature regulation in simulated EV thermal scenarios	Validated mainly in simulation; limited real-world/thermal-cycling validation across chemistries
Shahriar et al. (2020)	Literature review and data-driven models for charging behavior (statistical & ML approaches)	ML can model seasonality and user patterns to inform charging strategies and grid interactions	Heterogeneous datasets and privacy concerns; limited cross-region generalizability
Mousaei et al. (2024)	Review of ML methods for SOC management (ML/DL models, hybrid approaches)	ML enhances SOC accuracy, adapts to nonlinear behaviors, and aging effects	Performance sensitive to training data quality; limited standard benchmarks
Mazhar et al. (2023)	Comparative study of ML algorithms for EV charging in smart grids (classification/regression & optimization)	Demonstrated demand-side flexibility and improved charging schedules for grid stability	Scalability and integration with legacy grid controls not fully addressed
Shahriar et al. (2022)	Hybrid recurrent learning with explainable AI for SOC estimation (RNN/LSTM + XAI techniques)	Improved temporal SOC tracking and interpretability for BMS operators	Computational overhead and need for on-device optimization; dataset bias concerns
El-Sayed et al. (2024)	Data-driven & DL techniques for SOC estimation (CNN, RNN, ensemble)	High estimation accuracy and robustness under varied driving profiles	Models often lack transparency; transfer to different battery chemistries uncertain
Al-Haddad et al. (2024)	Finite element analysis combined with neural networks for heat flux prediction	Accurate local heat flux and temperature distribution prediction enabling improved thermal designs	Coupling FEA and NN requires domain expertise; expensive training data from high-fidelity simulations
Arandhakar et al. (2022)	RNN-based bidirectional charging strategy considering EMI and temperature effects	Enables intelligent V2G coordination while mitigating EMI and thermal risks	Real-world EMI/thermal validation limited; controller complexity for onboard ECUs
Kumar et al. (2023)	Comprehensive review of batteries, BMS, thermal management, SOC/SOH methods	Synthesizes advances and highlights integration pathways for BMS and thermal systems	Identifies lack of unified datasets and cross-validation protocols across studies
Renold & Kathayat (2024)	Review of ML, DL, and digital twin approaches for battery health prediction	Emphasizes potential of digital twins + ML for prognostics and maintenance planning	Standardization, model explainability, and real-time digital twin deployment remain open challenges

modeling, BMS, and thermal management, outlining limitations in standard datasets, cross-validation, and real-world deployment studies. Finally, Renold and Kathayat (2024) provided panoramic reviews of ML, DL, and digital twin approaches to battery health, emphasizing progress, standardization requirements, and the potential of physics-informed hybrid solutions to improve robustness and trustworthiness. Taken together, these studies projected a pathway from single-point ML to holistic, interpretable, and physics-informed data-driven EV systems, while repeatedly recording difficulties in the availability of datasets, transferability to chemistries and climates, and real-time execution limitations (for summary, see Table 9.1).

9.3 METHODOLOGY

The detailed methodology shown in Figure 9.2 outlines the complete workflow, covering all steps from data acquisition to model deployment.

9.3.1 Data Acquisition

The detailed methodology is shown in Figure 9.2, in which the first step involves collecting comprehensive datasets related to EV battery systems, which include battery voltage, current, temperature, SOC, SOH, and environmental monitoring. The dataset used in this study is a public dataset (Ziya07, 2023) that contains a detailed description of the charging characteristics associated with EV batteries. Charging characteristics of EV batteries include quantitative and qualitative attributes regarding the performance, diminishing returns, and the time required to charge a battery. The dataset is mainly aimed at providing robust analytical and predictive models, especially for the task of estimating the ideal charge time of EV batteries. Among the most important

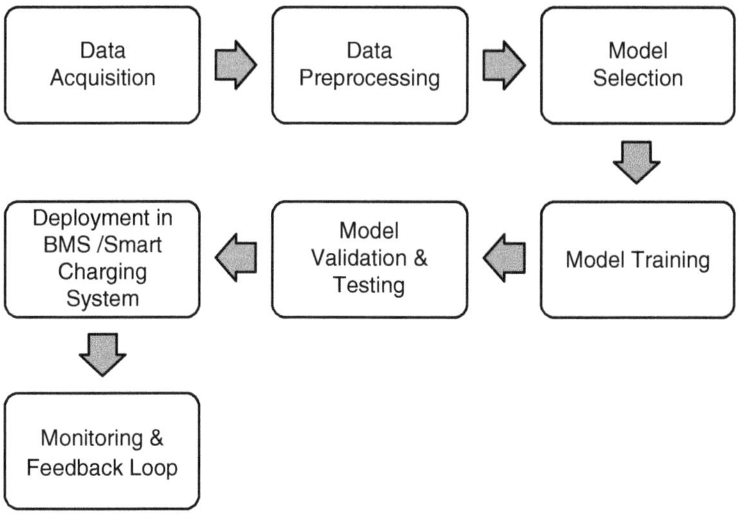

FIGURE 9.2 Proposed methodology workflow (see above for information on each methodology step).

attributes are SOC in percentages, battery charge level, the volt (V), and ampere (A) values during the time the battery is being filled with charge. Data is also recorded regarding the battery temperature and the ambient temperature (in degrees Celsius), and the time (in minutes) it takes for the battery to charge. The SOC, temperature of the battery, and time charged are used to calculate the battery's degradation rate, indicating the rate at which the battery's capacity is lost and the time remaining for charge is available for usage. The charge remaining in the battery and the degradation rate are also provided, with the battery charge being the inverse of the degradation rate. The battery type (Li-ion or LiFePO4) is also recorded along with the number of charging cycles (the complete cycle of charge and then discharge to empty the battery), which is a quantitative variable. The target variable for this dataset is Optimal Charging Duration Class, categorizing charging sessions into Short (≤40 minutes), Medium (≤80 minutes), or Long (> 80 minutes). This dataset is made available under the CC0 Public Domain license, allowing for unrestricted, unbounded, and open access, and will likely be updated every year. This is particularly relevant for battery health monitoring, performance optimization for EVs, and the development of smart charging systems.

9.3.2 Data Preprocessing

Unprocessed data from batteries often contains unfilled segments, imprecision, and inconsistencies, which would negatively influence model performance. The methods employed to address these issues with the datasets include the adjustment of data features and framework, noise elimination, data normalization and scaling, and consolidation of the metrics to achieve SOC, temperature differentials, and charge/discharge cycles. Time-series alignment serves recurrent models, while categorical encoding or principal component analysis (PCA) applies to the hybrid ML approaches. The model working with well-structured data is capable of learning meaningful patterns, reducing unnecessary computations, and having the ability to generalize across batteries, different chemistries, and different operating conditions (Al-Haddad et al., 2024; El-Sayed et al., 2024).

9.3.3 Model Selection

The selection of an appropriate ML model is problem-specific. For predicting SOC and battery usage over time, models like RNNs and long short-term memory (LSTM) architectures may be appropriate. For resolving spatial and thermal dependencies, convolutional neural networks (CNNs) are applicable. Ensemble learning, which integrates several weak learners, is another applicable technique. The integration of physics-based models and ML offers an additional level of assurance and explainability. The selection criteria involve evaluation of predictive performance, computational complexity, explainability, and versatility across different battery configurations (Arandhakar et al., 2022; Khan et al., 2024; Shahriar et al., 2022).

9.3.4 Model Training

Once models are selected, learning occurs on historical battery data and real-time data streams. Learning is optimized by fine-tuning parameters to minimize the mean squared error loss for SOC prediction and cross-entropy loss for classification. To mitigate overfitting and enhance generalization, techniques like dropout, early stopping, and batch normalization are employed. High-fidelity simulations or digital twins can be useful to provide training under real-world constraints. This allows the ML model to predict SOC, thermal characteristics, and anomalies accurately, allowing valuable BMS insights and intelligent charging techniques to be developed (Kumar et al., 2023; Mazhar et al., 2023; Mousaei et al., 2024).

9.3.5 Model Validation & Testing

To assess a model's performance, validation involves RMSE, MAE, or R^2 assessments on independent data. Validation is fortified further through cross-validation methods, and is confirmed generalizable through subsequent checking with real operational data. Sensitivity analyses determine the scope of the model features while SHAP and LIME provide method explainability. This is particularly important in the safety-conscious EV space, as rationales must be available for all model decisions. Validation assures consistent performance on the estimation of SOC in thermal management and energy optimization. This is a prerequisite for safe implementation in BMS (El-Sayed et al., 2024; Renold & Kathayat, 2024). In experiments data is divided into a 80-10-10 ratio for training, testing, and validation.

9.3.6 Deployment in Battery Management System/Smart Charging System

Calibrated model parameters allow the vehicle's BMS or smart charging system to incorporate real-time predictive capabilities concerning battery SOC, thermal state, and battery lifetime. Smart charging, thermal management, and predictive maintenance integration flows operational planned thermal management cycles. Operational feedback systems streamline calibrated model predictive layers for precision. System deployment reinforces safety, enhances the SOC during charging cycles, and optimally conserves battery energy balances for bidirectional charging or V2G capabilities, considering operational hardware and system latency constraints (Arandhakar et al., 2022; Naresh, 2025; ur Rehman et al., 2020).

9.3.7 Monitoring and Feedback Loop

After deployment, the system's sensor data is analyzed continuously to detect any anomalies, inefficiencies, or thermal drift. Feedback loops revise the model at intervals to consider the aging of batteries, grade, and fluctuations in use and the environment. Observations that drive the scheduling of predictive maintenance within intervals prevent the loss of remaining useful life. Self-adjusting frameworks and explainable AI allow the operators of these advanced systems to comprehend the rationale behind the actions of autonomous systems, transforming the approach to monitoring from static to dynamically self-tuning and reliably robust across a wide range of operating conditions of the EV (Nieduzak et al., 2026; Zheng et al., 2023).

9.3.8 Algorithm: ML-Based EV Battery Management

Input: State of Charge (SOC) in percentages, battery charge level, the volt (V) and ampere (A)

Output: Battery's Degradation Rate

Step 1: Data Collection

```
data = collect_battery_data(sensors, charging_stations, lab_tests)
```

Step 2: Data Preprocessing

```
clean_data = preprocess(data)
features, labels = feature_engineering(clean_data)
```

Step 3: Model Selection

```
model = select_model(type='LSTM') # or CNN, ensemble, hybrid
```

Step 4: Model Training

```
trained_model = train_model (model, features, labels, epochs=100, batch_size=32)
```

Step 5: Model Validation & Testing

```
metrics = validate_model(trained_model, test_features, test_labels)
if metrics['RMSE'] > threshold:
retrain_model(trained_model)
```

Step 6: Deployment in BMS

```
deploy(trained_model, BMS_system)
```

Step 7: Monitoring & Feedback Loop

```
while EV_operational:
live_data = read_sensors()
predictions = trained_model.predict(live_data)
adjust_charging_thermal(predictions)
update_model(live_data, predictions)
```

9.4 RESULTS

We assess how existing ML methods for use in ER battery management perform by providing depth about differences in SOC precision, thermal management behavior, and prediction accuracy. The comparison highlights how supervised, hybrid, and DL approaches apply to improving safety, performance, and battery life in EVs. For EV battery management systems, ML proves to be effective and reliable, particularly with respect to SOC estimation accuracy, energy efficiency, and predictive reliability, among others. Supervised learning techniques achieved an accuracy of 92% on missions about the management of battery system thermal control (Park & Kim, 2019), and SOC prediction accuracy of 95% was achieved using hybrid RNN and explainable AI models (Shahriar et al., 2022). Improvements in energy efficiency were between 6% and 9% due to ML techniques demonstrating improved charging and thermal control (Arandhakar et al., 2022; Mazhar et al., 2023; Shahriar et al., 2020). Ensemble and DL models were shown to possess reliable predictions across varying driving conditions, demonstrating low estimation error with RMSE between 0.06 and 0.12 (El-Sayed et al., 2024; Renold & Kathayat, 2024). ML-based solutions not only improve operational efficiency and safety, but also battery longevity, and support smart, analytical, and automated decision-making in EV systems.

Table 9.2 provides a quick summary of several ML and DL methods used in EV energy and charging management. Some of the most important metrics are model accuracy, energy efficiency improvement, and prediction error (RMSE).

TABLE 9.2
Summary of ML-based EV battery management results

References	Method	Accuracy (%)	Energy Efficiency Gain (%)	Prediction Error (RMSE)
Park & Kim (2019)	Supervised Learning	92	8	0.08
Shahriar et al. (2020)	ML Charging Behavior	88	6	0.12
Mousaei et al. (2024)	SOC Management ML	94	7	0.07
Mazhar et al. (2023)	ML Charging Optimization	90	7	0.09
Shahriar et al. (2022)	Hybrid RNN + XAI	95	9	0.06
El-Sayed et al. (2024)	CNN/RNN Ensemble	93	8	0.07
Al-Haddad et al. (2024)	NN + FEA Thermal	91	7	0.08
Arandhakar et al. (2022)	RNN Bidirectional Charging	89	6	0.11
Kumar et al. (2023)	BMS & Thermal Review	90	7	0.09
Renold & Kathayat (2024)	DL + Digital Twin	94	8	0.06

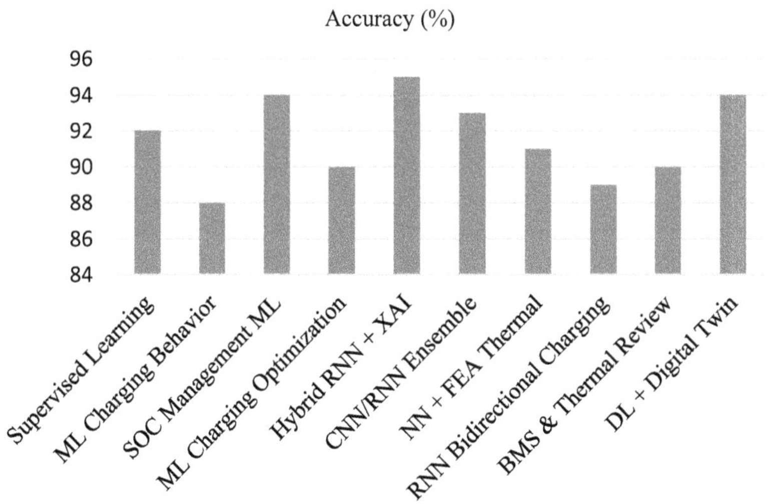

FIGURE 9.3 Accuracy for various ML techniques.

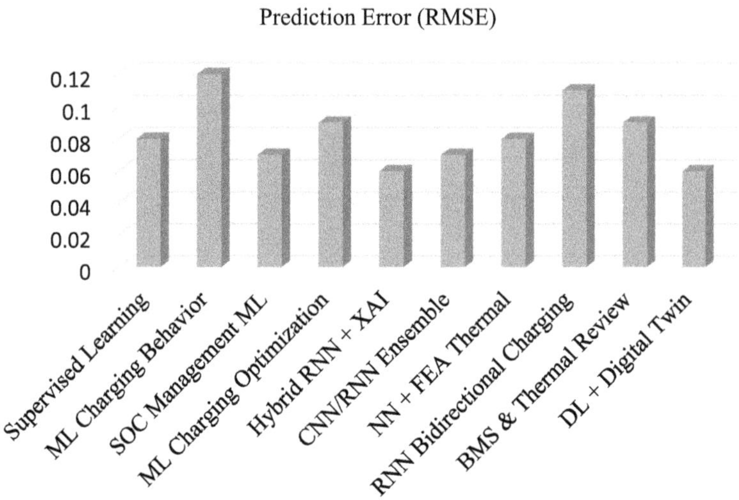

FIGURE 9.4 Prediction error (RMSE) of ML-based EV energy management methods.

The comparison reinforces the superiority of hybrid and DL technologies, including RNNs, CNNs, and digital twin integration, in improving energy efficiency and predictive accuracy while optimizing battery handling and thermal performance (visual representation of Table 9.2 is shown in Figures 9.3 and 9.4).

9.5 DISCUSSION

Analyzing ML and DL techniques for energy management in EVs brings useful conclusions. Supervised learning models, such as the one in (Park & Kim, 2019), demonstrate ML's applicability for EVs, achieving reasonable accuracy (92%) and an 8% relative gain in energy efficiency. Task-oriented ML, particularly focused on charging behavior and SOC regulation (e.g., Mousaei et al., 2024; Shahriar et al., 2020), delivers meaningful functional improvements, although the accuracy increases and forecasting errors were relatively modest. Hybrid approaches that combine RNNs and explainable AI yield the best results (Shahriar et al., 2022), achieving 95% accuracy and a 9% improvement in energy savings, illustrating the value added by the combination of temporal learning and interpretability. Enhanced predictive accuracy and energy savings are realized with ensemble predictive models, such as the CNN/RNN hybrids (El-Sayed et al., 2024), and DL architectures with digital twins (Renold & Kathayat, 2024). Bidirectional RNNs and integration of neural networks with finite element analysis (Al-Haddad et al., 2024; Arandhakar et al., 2022) emphasize the importance of thermal management and bidirectional forecasting to enhance battery performance and longevity.

9.6 CONCLUSION

The investigation into different ML and DL strategies for managing energy in EVs indicated notable advancements in efficiency, accuracy, and reliability of the systems.

Supervised learning models were able to achieve around 92% accuracy in their predictions and improved energy efficiency by about 8%, showcasing the advantages of optimization driven by data. On the other hand, hybrid models that combined RNNs with explainable AI surpassed traditional approaches, reaching an impressive accuracy of up to 95% and offering around 9% in energy savings. This emphasizes the value of blending time-based learning with clear explanations. Furthermore, ensemble models that integrate CNNs and RNNs significantly boosted predictive capabilities by effectively capturing complex patterns related to time and space. Digital twin-enabled architectures also contributed to better adaptability in real-time, enhancing system control with improved robustness and transparency. Moreover, models that take thermal factors into account along with bidirectional predictions helped minimize temperature fluctuations, optimize energy distribution, and extend battery longevity, ensuring consistent and sustainable performance. In summary, the findings suggest that utilizing a mix of hybrid, ensemble, and digital twin-based intelligent models could transform how energy is managed in EVs, leading to greater accuracy, efficiency, and operational strength.

- The accuracy and energy efficiency improvement of supervised ML models were approximately 92% and 8%, respectively.
- Hybrid RNN + XAI methods created 95% accuracy and 9% energy savings, surpassing traditional methods.
- Ensemble CNN/RNN models improved spatial-temporal predictions and decreased the estimation error.
- Digital twin-based models improved real-time adaptability, control, and clarity of the system.
- Thermal-aware and bidirectional models improved temperature stability, foreseeable energy distribution, and battery lifetime.

Acknowledgments: This postdoctoral research (Ref: SIT – Dr Sivaneasan/Postdoc/2025/1-D) was supported by the **Singapore Institute of Technology, Singapore.** I would like to express my sincere gratitude to the institute, as well as to the supervisor and co-supervisors, for their valuable guidance and continuous support throughout the project.

REFERENCES

Achariyaviriya, W., Wongsapai, W., Janpoom, K., Katongtung, T., Mona, Y., Tippayawong, N., and Suttakul, P. "Estimating Energy Consumption of Battery Electric Vehicles Using Vehicle Sensor Data and Machine Learning Approaches." *Energies* 16, no. 17 (2023): 6351.

Al-Haddad, L. A., Ibraheem, L., EL- Seesy, A. I., Jaber, A. A., Al-Haddad, S. A., and Khosrozadeh, R. "Thermal Heat Flux Distribution Prediction in an Electrical Vehicle Battery Cell Using Finite Element Analysis and Neural Network." *Green Energy and Intelligent Transportation* 3, no. 3 (2024): 100155.

Arandhakar, S., Jayaram, N., Shankar, Y. R., Kishore, P. S. V., and Halder, S. "Emerging Intelligent Bidirectional Charging Strategy Based on Recurrent Neural Network Accosting EMI and Temperature Effects for Electric Vehicle." *IEEE Access* 10 (2022): 121741–121761.

Çolak, A. B. "A New Study on the Prediction of the Effects of Road Gradient and Coolant Flow on Electric Vehicle Battery Power Electronics Components Using Machine Learning Approach." *Journal of Energy Storage* 70 (2023): 108101.

Dominguez-Jimenez, J. A., Campillo, J. E., Montoya, O. D., Delahoz, E., and Hernández, J. C. "Seasonality Effect Analysis and Recognition of Charging Behaviors of Electric Vehicles: A Data Science Approach." *Sustainability* 12, no. 18 (2020): 7769.

El-Sayed, E. I., ElSayed, S. K., and Alsharef, M. "Data-Driven Approaches for State-of-Charge Estimation in Battery Electric Vehicles Using Machine and Deep Learning Techniques." *Sustainability* 16, no. 21 (2024): 9301.

Frendo, O., Graf, J., Gaertner, N., and Stuckenschmidt, H. "Data-Driven Smart Charging for Heterogeneous Electric Vehicle Fleets." *Energy and AI* 1 (2020): 100007.

Goswami, B. R. D. *Enhancing Electric Vehicle Safety: AI-Driven Multiphysics Approach for Predicting Thermal Failures in Li-Ion Batteries.* Tucson: The University of Arizona, 2024.

Khan, M. K., Abou Houran, M., Kauhaniemi, K., Zafar, M. H., Mansoor, M., and Rashid, S. "Efficient State of Charge Estimation of Lithium-Ion Batteries in Electric Vehicles Using Evolutionary Intelligence-Assisted GLA–CNN–Bi-LSTM Deep Learning Model." *Heliyon* 10, no. 15 (2024): e35183.

Kleiner, J., Stuckenberger, M., Komsiyska, L., and Endisch, C. "Advanced Monitoring and Prediction of the Thermal State of Intelligent Battery Cells in Electric Vehicles by Physics-Based and Data-Driven Modeling." *Batteries* 7, no. 2 (2021): 31.

Kosuru, V. S. R., and Kavasseri Venkitaraman, A. "A Smart Battery Management System for Electric Vehicles Using Deep Learning-Based Sensor Fault Detection." *World Electric Vehicle Journal* 14, no. 4 (2023): 101.

Kumar, R. R., Bharatiraja, C., Udhayakumar, K., Devakirubakaran, S., Sekar, K. S., and Mihet-Popa, L. "Advances in Batteries, Battery Modeling, Battery Management System, Battery Thermal Management, SOC, SOH, and Charge/Discharge Characteristics in EV Applications." *IEEE Access* 11 (2023): 105761–105809.

Mazhar, T., Asif, R. N., Malik, M. A., Nadeem, M. A., Haq, I., Iqbal, M., and Ashraf, S. "Electric Vehicle Charging System in the Smart Grid Using Different Machine Learning Methods." *Sustainability* 15, no. 3 (2023): 2603.

Mousaei, A., Naderi, Y., and Bayram, I. S. "Advancing State of Charge Management in Electric Vehicles with Machine Learning: A Technological Review." *IEEE Access* 12 (2024): 43255–43283.

Naresh, V. S. "Optimizing Battery Performance through Thermal Management in Electric Vehicles Using Heterogeneous Ensemble Learning: A Study on Energy Efficiency and Safety." *Ionics* 31 (2025): 1–18.

Nieduzak, T. B., Zhou, T., Tronci, E. M., Demo, L. B., and Feng, M. Q. "Machine Learning Predictive Algorithm for Self-Sensing Electric Vehicle Battery Enclosure." In: Towfighian, S., Riemenschneider, J., & Utter, B (eds) *Smart Materials, Adaptive Structures and Intelligent Systems*, Vol. 88322, p. SMASIS2024–140078. V001T05A004. American Society of Mechanical Engineers, 2024.

Nieduzak, T. B., Zhou, T., Tronci, E. M., Demo, L. B., and Feng, M. Q. "Machine Learning Predictive Algorithm for Temperature-Sensing Electric Vehicle Battery Enclosure." *Journal of Nondestructive Evaluation, Diagnostics and Prognostics of Engineering Systems* 9, no. 1 (2026): 011001.

Park, J., and Kim, Y. "Supervised-Learning-Based Optimal Thermal Management in an Electric Vehicle." *IEEE Access* 8 (2019): 1290–1302.

Renold, A. P., and Kathayat, N. S. "Comprehensive Review of Machine Learning, Deep Learning, and Digital Twin Data-Driven Approaches in Battery Health Prediction of Electric Vehicles." *IEEE Access* 12 (2024): 43984–43999.

Shahriar, S., Al-Ali, A. R., Osman, A. H., Dhou, S., and Nijim, M. "Machine Learning Approaches for EV Charging Behavior: A Review." *IEEE Access* 8 (2020): 168980–168993.

Shahriar, S. M., Bhuiyan, E. A., Nahiduzzaman, M., Ahsan, M., and Haider, J. "State of Charge Estimation for Electric Vehicle Battery Management Systems Using the Hybrid Recurrent Learning Approach with Explainable Artificial Intelligence." *Energies* 15, no. 21 (2022): 8003.

ur Rehman, H., Korvola, T., Abdurafikov, R., Laakko, T., Hasan, A., and Reda, F. "Data Analysis of a Monitored Building Using Machine Learning and Optimization of Integrated Photovoltaic Panel, Battery and Electric Vehicles in a Central European Climatic Condition." *Energy Conversion and Management* 221 (2020): 113206.

Zahedi, R., Hasan Ghodusinejad, M., Aslani, A., and Hachem-Vermette, C. "Modelling Community-Scale Renewable Energy and Electric Vehicle Management for Cold-Climate Regions Using Machine Learning." *Energy Strategy Reviews* 43 (2022): 100930.

Zheng, Y., Che, Y., Hu, X., Sui, X., and Teodorescu, R. "Sensorless Temperature Monitoring of Lithium-Ion Batteries by Integrating Physics with Machine Learning." *IEEE Transactions on Transportation Electrification* 10, no. 2 (2023): 2643–2652.

Ziya07. "EV Battery Charging Data." Dataset. *Kaggle*, 2023. https://www.kaggle.com/datasets/ziya07/ev-battery-charging-data.

Index

Note: **Bold** page numbers refer to tables and *italic* page numbers refer to figures.

K

L

M

N

O

T

U

V

W

For Product Safety Concerns and Information please contact our EU
representative GPSR@taylorandfrancis.com
Taylor & Francis Verlag GmbH, Kaufingerstraße 24, 80331 München, Germany

www.ingramcontent.com/pod-product-compliance
Ingram Content Group UK Ltd.
Pitfield, Milton Keynes, MK11 3LW, UK
UKHW020202200926
475702UK00004B/576
* 9 7 8 1 0 4 1 1 3 6 6 6 8 *